Bi

James Egan was born in 1985 and grew up in Portarlington, Co. Laois in the Midlands of Ireland.
In 2008, James moved to England and studied in Oxford.
James married his wife in 2012 and currently lives in Havant in Hampshire.
James had his first book, 365 Ways to Stop Sabotaging Your Life, published in 2014.
Several of James' books have become No.1 Best Sellers in the UK including 1000 Facts about Horror Movies, 3000 Facts About the Greatest Movies Ever, 365 Things People Believe That Aren't True, Another 365 Things People Believe That Aren't True, and 500 Things People Believe That Aren't True.

Books by James Egan

Fairytale
Inherit the Earth
Inherit the Earth: The Animal Kingdom
1000 Facts About the United States
Words That Need to Exist in English
Hilarious Things That Kids Say
Hilarious Things That Mums Say
1000 Facts about TV Shows Vol. 1-3
1000 Facts about Animated Shows Vol. 1-3
1000 Facts about Actors Vol. 1-3
1000 Facts about Countries Vol. 1-3
Dinosaurs Had Feathers (and other Random Facts)
1000 Facts about Animals Vol. 1-3
1000 Facts about James Bond
1000 Inspiring Facts
How to Psychologically Survive Cancer
1000 Out-of-this-World Facts about Space
1000 Facts about the Greatest Movies Ever Vol. 1-3
1000 Facts about Film Directors
1000 Facts about Superhero Movies Vol. 1-3
1000 Facts about Superheroes Vol. 1-3
1000 Facts about Supervillains Vol. 1-3
1000 Facts about Comic Books Vol. 1-3
1000 Facts about Animated Films Vol. 1-3
1000 Facts about Horror Movies Vol. 1-3
1000 Facts about American Presidents
Adorable Animal Facts
1000 Facts about Video Games Vol. 1-3
Things People Believe That Aren't True Vol. 1-4
1000 Fact about Film Director
The Mega Misconception Book
3000 Astounding Quotes
1000 Facts About Comic Book Characters Vol. 1-3
100 Classic Stories in 100 Pages
500 Facts about Godzilla
365 Ways to Stop Sabotaging Your Life
Flat Earthers Around the Globe
1000 Facts about Historic Figures Vol. 1-3
1000 Facts About Writers
1000 Facts about Ireland
The Biggest Movie Plotholes
1000 Facts about the Human Body

1000 Out-of-This-World Facts About Space

By

James Egan

ISBN: 9781326467234

Lulu Publishing Services rev. date: 02/11/2015

Dedicated to
my astronomy-loving pioneer,
Flora

And

Peter Bainbridge-Clayton

Content

Planets

1. A chain of craters on a planet's surface is known as a Catena.

2. A broken terrain on a planet's surface is called a Chaos.

3. A canyon on a planet's surface is known as a Chasma.

4. If a planet has a closer orbit to the Sun than Earth (Mercury and Venus) it is called an Inferior Planet. If a planet has a broader orbit to the Sun than Earth, it known as a Superior Planet.

5. Many people believe that planets have a circular orbit around their star. This is not true. All planets orbit their neighbouring star in an ellipse (oval-shape.) This means that it varies how far a planet can be from its star. This is known as an elliptical orbit.

6. When a planet is at its closest point to the Sun, it is known as perihelion. When a planet is at its furthest point from the Sun, it is known as aphelion.

7. There is no true definition of the word "planet." It doesn't have to be round. Most planets aren't round (including Earth.) Some moons and asteroids are rounder than Earth.

   A planet doesn't need to have a moon. Mercury and Venus don't. However, some asteroids have moons.

   Pluto isn't a planet but it has a moon. How does that work?

   Does a planet have to be a certain size? Nope. Jupiter's moon, Ganymede, is bigger than Mercury. So why is Ganymede a moon but Mercury is a planet?

   Sadly, the word "planet" can't be defined.

8. The word "planet" is Greek for "wanderer."

9. Scientists have recently discovered a new kind of planetoid called a Chthonian planet. This is when a Gas Giant (like Jupiter) gets too close to its neighbouring Sun, causing the gas to evaporate until only a core remains.

10. Planets and moons makes sounds! How is this possible if there is no sound in space? Well, planets and moons can create sounds

from interactions of charged particles from the solar wind, ionosphere, and planetary magnetosphere.

...If you don't understand any of that, the alternative explanation is... magic.

Mercury

11. If you could physically stand on Mercury, the Sun would look like it was the size of a Ping-Pong ball held at arm's length.

12. One day on Mercury lasts 1,408 hours – the same as how long one Monday on Earth feels.

13. Mercury has the most elliptical orbit around the Sun. This means that Mercury's distance from the Sun varies the most. Mercury is usually 36 million miles from the Sun. At its closest, Mercury is 28 million miles from the Sun. At its farthest, it's 43 million miles from the Sun.

14. When Mercury is at its closest to the Sun, it receives twice as much light and heat as it gets when it's at its farthest point.

15. The planet is 3,044 miles in diameter. This makes Mercury 1/3rd the width of Earth.

16. A year on Mercury is as long as 88 days on Earth.

17. Mercury has no moons.

18. Even though Mercury is incredibly hot, there is ice there. The ice is buried deep beneath the planet's caverns where sunlight can't get to it. It's been left over from ice comets that have collided into the planet over the years. Most ice is evaporated instantly but traces of it land in spots where the Sun doesn't reach it. Over millions of years, Mercury has accumulated billions of tons of ice.

19. If a planet has an area that sunlight never reaches, it's called a Cold Trap. All of Mercury's ice is hiding in Cold Traps.

20. Mercury is covered in hundreds of craters, making it the most cratered planet in the Solar System. The biggest crater is Caloris Planitia, measuring 960 miles wide.

All of Mercury's craters are named after artists. Some of the craters include Bach, Beckett, Chekhov, Dostoevsky, Goya, Goethe, Horace, Ibsen, Imhotep, Keats, Kuiper, Lennon, Twain, Melville, Milton, Moliere, Mozart, Ovid, Picasso, Poe, Raphael, Shakespeare, Tolstoy, Wagner, Yeats,

Hemingway, Botticelli, Tolkien, Beethoven, Dali, Bronte, Byron, and Dickens.

21. Mercury can be up to 430 degrees Celsius during the day. At night, Mercury's surface is -170 degrees Celsius.

22. Mercury has the biggest core in proportion to the planet's size of any planet in the Solar System. Its core is 3/4th the size of the planet itself.

23. Mercury moves 29 miles per second. That's 107,372 miles per hour.

24. Mercury has almost no atmosphere.

25. Mercury has 38% the gravity of Earth. Weirdly, Mars also has 38% the gravity of Earth.

26. Mercury is the second densest planet in the Solar System.

27. Mercury was first recorded in 1400 BC by astronomers from Assyria (Ancient Ethiopia.) However, the Assyrians didn't think it was a planet. They just noticed a light in the sky that happened to be Mercury.

28. Mercury is the smallest planet in the Solar System. (It used to be Pluto before he got kicked out.)

29. At its closest, Mercury is 48 million miles away from Earth.

30. For a long time, astronomers couldn't figure out how long Mercury's day was because it was too small and far away for astronomers to observe surface changes.

31. Only two satellites have visited Mercury – Mariner 10 in 1974 and Messenger in 2011. Messenger took pictures of the planet's mile-deep pits.

32. Every 12 years or so, Mercury can be seen crossing the Sun.

33. Mercury's surface looks like it has wrinkles. These wrinkles are known as Lobate Scarps. They were formed as the iron core of the planet cooled and contracted over millions of years.

34. Mercury can only be seen after the Sun has risen and before the Sun has set.

35. Mercury is a hundred times closer to the Sun than Pluto.

36. When one year has passed on Pluto, 1,028 years have passed on Mercury.

37. It takes 3.2 minutes for the Sun's light to reach Mercury.

38. Mercury is named after the Roman Messenger of the Gods.

39. Mercury was the only planet whose motion couldn't be fully explained by Isaac Newton's equation. The planet feels so much gravity from the Sun, that it needed Einstein's more accurate General Theory of Relativity to explain its orbit.

Venus

40. On Venus, a day is longer than a year. One day on Venus is 243 Earth days. One year on Venus is 224 Earth days. Venus has the longest day of any planet in the Solar System.

41. Venus rotates so slowly that you could jog at the equator and lap the planet before it rotated once.

42. Venus is flipped upside-down. Its top point is its South Pole and vice versa. Because of this, Venus spins in the opposite direction to the other planets in the Solar System. This is known as retrograde motion.

43. Venus probably had water millions of years ago but it evaporated over time.

44. Venus is the most spherical of all the planets in the Solar System. The other planets tend to be flat at the top and bottom and bulge in the middle so they are not actually spherical.

45. Venus' atmosphere has a rainbow.

46. There are no tectonic plates on the planet.

47. Nearly all of Venus' volcanoes have gone extinct.

48. Venus has the same temperature on every part of the surface.

49. Venus is the third brightest thing in the sky, after the Sun and the Moon. Venus is so bright, that it can be seen during the day. The best time to see Venus is before sunset or after sunrise.

50. Venus has the least elliptical orbit around the Sun in the Solar System. At its closest, Venus is 66 million miles away from the Sun. At its furthest, Venus is 68 million miles from the Sun.

51. At its closest, Venus is 25 million miles away from Earth.

52. Venus is regularly reported as a UFO.

53. It's so hot on Venus that the sulphuric acid rain evaporates before it hits the ground.

54. Venus' diameter is 7,521 miles.

55. Because Venus looks so beautiful in the sky, it was named after the Roman Goddess of beauty.

56. The Mayans were the first civilization to record observations of Venus in 650 AD.

57. Venus is the most similar to Earth in size. Venus has 95% of the diameter of Earth and 80% of its mass. However, Venus has massive differences to Earth as well. Like...

58. Venus is the hottest planet in the Solar System. Its surface temperature is 460 degrees Celsius. That's hot enough to melt lead. Or people.

59. It snows metal on Venus.

60. Its atmosphere is composed of 95% carbon dioxide.

61. Venus doesn't have a magnetic field so it has no protection from the solar wind. This is the main reason why the planet is so hot.

62. Venus’ atmosphere pressure is 92 times higher than Earth’s.

63. Venera 7 was the first space probe to land on another planet. It was built by the Soviets and was launched on August 17th, 1970. It landed on Venus shortly after. It took a single photograph of the planet’s surface. Within two hours, the robot was crushed and melted by the planet’s atmosphere.

64. Every surface feature on Venus (mountain, crater, volcano, etc.) is named after a famous woman or goddess. Some of these include Abeona Mons, Adivar, Akna Montes, Barton, Cleopatra, Ciuacoatl Mons, Danu Montes, Fand Mons, Freyja Montes, Isabella, Irnini Mons, Maat Mons, Mariko, Mona Lisa, Renpet Mons, Ruth, Sapas Mons, Stefania, Sif Mons, Theia Mons and Ushas Mons.

65. Venus looks so different in the morning and evening, that early civilizations thought it was two different stars. The Greeks called it Phosphorus and Hesperus. The Romans called the planet Lucifer and Vesper.

66. Venus moves 30 miles per second.

67. It takes exactly six minutes for the Sun's light to reach Venus.

68. For years, the dense clouds on Venus' surface made it impossible for astronomers to see what kind of world it was. Scientists imagined it was a swamp-like paradise populated with Venusian aliens. Ironically, it is the most uninhabitable planet in the Solar System.

69. Although Venus is considered inhospitable, the conditions are surprisingly similar to Earth 30 miles above the surface. Unlike Mars, Venus' atmosphere can shield us from the Sun's ultraviolet rays. Scientists have considered a possible sky city in the future hovering over Venus.

Earth

70. Earth is the only planet in the Solar System that is not named after a Greek or Roman god.

71. The Earth is 4.5 billion years old.

72. If you compressed the last 4.5 billion years into one single day, humanity only existed at 23 hours, 59 minutes and 12 seconds. On this scale, mankind has only been around for 48 seconds.

73. On average, Earth is 93 million miles away from the Sun. At its closest, Earth is 91 million miles from the Sun. At its furthest, Earth is 94 million miles from the Sun.

74. Earth has about 8.7 million different types of life forms. This is a bit more than the other planets that have approximately zero.

75. Some people have made the observation, "If our world is made of 71% water, why is it called Earth?"

However, Earth is made of 1% water. 71% of the Earth's *surface* is made of water.

Most of Earth is made of a calcium titanate material called perovskite.

76. When Earth was first forming, it was a big ball of rock and lava. This was known as the Hadean Era.

77. Earth was called Gaea by the Ancient Greeks. Gaea was the goddess of the world.

78. Terra is what the Ancient Romans called the Earth.

79. Earth is the densest planet in the Solar System.

80. Earth has this weird meteorological phenomenon caused by reflection, refraction, and dispersion of light in water droplets, which creates a spectral arc of light in the sky. Most people call it a rainbow.

81. The Earth is 7,926 miles in diameter.

82. Life on Earth began between 3.7-4.4 billion years ago.

83. Light travels around the Earth in $1/10^{th}$ of a second.

84. It travels from the Earth to the Moon in 1.3 seconds.

85. It takes 8.3 minutes for the Sun's light to reach Earth.

86. Earth weights six sextillion tons. That's 6,000,000,000,000,000,000,000 tons.

87. 252 million years ago, water on Earth was pink.

88. Many people know that millions of years ago, the continents of Earth were merged together into a supercontinent called Pangaea. But this was only one supercontinent. A supercontinent forms, then splits apart millions of years later and reforms into a new supercontinent millions of years after that. So far, there have been seven supercontinents on Earth – Vaalbra, Ur, Kenorland, Columbia, Rodinia, Pannotia and Pangaea. The next supercontinent will be Ultima. It will form in 100 million years.

89. The hottest natural temperature ever recorded on Earth was 70.7 degrees Celsius in an Iranian desert. The coldest natural

temperature recorded on Earth was -89.2 degrees Celsius in Antarctica.

90. Earth is the largest solid planet in the Solar System.

91. Life on Earth is possible because our planet has Goldilocks Conditions. This means that Earth is in the right position away from the Sun so it isn't too hot or too cold; it's just right for life to thrive.

92. Many people believe that the Greek god, Atlas, held the Earth up on his shoulders. This is a misconception. In Greek mythology, Atlas held the Heavens on his back.

93. At this moment, you are 3,976 miles from the centre of the Earth.

94. The Earth moves 24 miles per second. That's 67,000 miles per hour.

95. The Earth's mantle is 1,800 miles thick.

96. Most people think the mantle's composition is like lava but geologists say it's more like plastic... if it was really, really, really hot.

97. In the centre of the Earth is the core. The core is 5,500 degrees Celsius.

98. The reason why the core is so hot is because of decaying radioactive elements such as uranium, friction, gravity, and the leftover heat from the Earth's formation.

99. The core of the Earth is made of iron and nickel. Even though the core is 5,500 degrees, the iron is solid, not liquid.

100. Although it's common knowledge that the Earth's top layer is the crust, most people don't realise that there are two types of crust. Underneath the oceans is the Oceanic Crust, measuring three miles thick.

Underneath land is the Continental Crust, which is about 25 miles thick.

101. As many as ten million pieces of human-made debris are estimated to be in Earth's orbit.

102. Earth is not round. Earth is flat at the poles and bulges at the equator. The technical term for its shape is "oblate spheroid."

103. Over 200 objects were released into Earth's orbit by the Mir space station during its first ten years of operation.

104. In 2005, at least 13 nuclear reactor fuel cores, eight thermoelectric generators, and 32 nuclear reactors were known to be in Earth's orbit.

105. The oldest debris that is still in Earth's orbit is Vanguard I. It was launched in 1958 and was the US' second satellite.

106. During the first American spacewalk in 1965, Gemini 4 astronaut, Edward White, lost a glove. For a month, the glove stayed in orbit, travelling at 17,398 miles per hour.

107. There is a post office 200 miles above the Earth. It even has a zipcode (just in case it gets mixed up with all those other space post offices.) Its address is China Post Space Office, Space.

108. Many assume that Earth is closest to the Sun during the summer. This isn't true. Bizarrely, Earth is closest to the Sun in January.

109. The iconic photograph of the Earth taken from the Apollo 17 mission is known as The Blue Marble.

110. Continental drift makes the continents move a few centimetres per year. It will take about 60,000 years for a continent to move one mile.

111. Earth's atmosphere is composed of 78% nitrogen, 21% oxygen, nearly 1% argon, and a very small percentage is made up of trace gases. Carbon dioxide only makes up 0.04% of the Earth's atmosphere.

112. Although there is no exact point where Earth's atmosphere stops and space begins, out of convenience, it is said that space starts 62 miles up from Earth's surface. It is called the Karman Line.

113. The Ozone Layer starts 15 miles above Earth's surface. The Ozone absorbs ultraviolet light; a light that can tear biological molecules apart.

114. The seasons of the year are not based on the distance between the Earth and the Sun.

The seasons are caused by the 23-degree axial tilt. We tilt towards the Sun in the summer and tilt away in the winter.

115. Earth's axis wobbles upon every rotation. Although this wobble is very slight, it will affect how we see the stars over the next few thousand years. For example, Polaris is commonly known as the North Star. However, thousands of years ago, the Ancient Egyptians saw Thuban as their North Star because it was in the same position that Polaris is in now. In 11,000 years, Vega will become Earth's North Star.

116. The Earth's rotation is slowing down. The deceleration is so slight that it is impossible to notice. But in 140 million years, the Earth will have 25 hours in a day.

117. 95% of the ocean depths have never been seen by human eyes.

118. Anaximander of Ancient Greece figured out that the Earth was not flat in 550 BC. He first considered this when he noticed that ships in water always seemed to disappear from the bottom-up. He believed that this

was only possible if the Earth's surface was curved, not flat.

119. Eratosthenes of Ancient Greece was the first person to work out how big the Earth was with simple geometry. He concluded that the Earth was 24,860 miles in circumference. Earth's circumference is 24,901 miles but it is astounding that he was so accurate considering he used basic mathematics to figure it out.

120. There is enough gold in the Earth's core to cover the world's surface to a depth of 1.5ft.

121. There is a bubble of bacteria 33,000ft above the planet's surface.

122. Okay…the big one. What are the Northern Lights?

The Northern Lights (also known as the Aurora Borealis) are visible in the North Pole and Iceland (and rarely in Scotland and Northern Ireland.) The Aurora is formed when the Earth's magnetic field traps solar wind particles and channels some of the particles into its atmosphere, where they smash into air molecules. This collision energises the molecules by emitting colourful lights. Nitrogen glows red and blue. Oxygen glows red and green.

The best time and place to see the Aurora Borealis is in Iceland from September to February between 12-3am. It can last for seconds, minutes, or hours.

Many people are unaware that there is an Aurora Australis (The Southern Lights) that is visible in New Zealand.

123. If you ever feel like your life is boring, remember this quote by writer, Douglas Adams, "The fact that we live at the bottom of a deep gravity well, on the surface of a gas-covered planet going around a nuclear fireball 90 million miles away and think this is normal is obviously some indication of how skewed our perspective tends to be."

124. If all human beings vanished overnight, the last man-made structures standing will be the Pyramids and Mount Rushmore. They are made of granite so they will resist erosion.

125. You could fit all the planets in the Solar System in-between Earth and the Moon.

126. No one has left low-Earth orbit since 1972.

127. An average person walks the equivalent of four times around the Earth during his or her lifetime.

128. The Earth's core is two-and-a-half-years younger than the crust. If you think it's because it formed later than the crust, it's much more complicated than that. According to a 2016 study, the core is so dense, that it has a higher gravitational field, which makes time pass more slowly.

Mars

129. Mars is named after the Roman god of War. The Ancient Romans believed that the planet Mars was the God of War.

130. Mars is the only planet inhabited solely by robots.

131. NASA sent The Opportunity rover robot to Mars in 2003. It was expected to last three months. It is still operational today.

132. The Opportunity has travelled over 26 miles on Mars.

133. Mars' diameter is 4,220 miles.

134. Mars is half the size of Earth.

135. On average, Mars is 142 million miles from the Sun. At its closest, Mars is 127 million miles from Sun.

136. At its furthest, Mars is 155 million miles from Earth. At its closest, Mars is 34 million miles from Earth.

137. A year on Mars is 687 Earth days.

138. One day on Mars lasts for 24 hours and 39 minutes.

139. Mars has polar ice caps.

140. The equator is the hottest part of Mars, reaching temperatures of 20 degrees Celsius. The poles have the coldest temperatures on Mars reaching temperatures of -153 degrees Celsius.

141. Despite the fact that it's nicknamed The Red Planet, Mars is butterscotch brown. If you have seen pictures of the red surface of Mars, those pictures were originally black-and-white and they were coloured in to appear how astronomers believed the surface looked.

142. On the surface of Mars, the Sun seems to be about the same size as a pea held at arms-length.

143. It takes 12.7 minutes for the Sun's light to reach Mars.

144. The Curiosity Robot on Mars sings Happy Birthday to itself once a year.

145. Every five years, the planet is blanketed by a dust storm that blocks out the Sun for several months.

146. Mars moves 20 miles per second.

147. Mars harbours a volcano called Olympus Mons. It is the biggest volcano (and mountain) in the Solar System, measuring nearly 16 miles high. That makes it almost three times higher than Mount Everest. What's even more impressive is that it is 370 miles wide. This means that Olympus Mons is slightly smaller… than France.

148. Mars' most identifiable feature is Valles Marineris. It is a gimungous crack on the planet's surface. It's 2,485 miles long, 124 miles wide, and four miles deep. That's ten times longer and wider than the Grand Canyon.

149. Mars has an atmosphere but it's only 1% as thick as Earth's atmosphere and it is mostly composed of carbon dioxide.

150. Mars has storms similar to tornados called dust-devils.

151. As of December 2014, Mars has the most active spacecraft orbiting a planet besides Earth. The spacecraft are called Odyssey, Express, Reconnaissance, MAVEN, and Mangalyaan.

152. Mars has two moons called Deimos and Phobos. Deimos is only 16 miles across. Phobos is merely nine miles across.

153. Mars' moons are shaped like potatoes. Because of Deimos and Phobos' odd shape, it's likely that they are not conventional moons and are simply asteroids that got dislodged from the nearby Asteroid Belt and got absorbed into Mars' gravity field.

154. Phobos rotates around Mars faster than the planet spins, meaning that it gradually gets closer and closer to the planet's surface. In a few million years, it will enter the planet's atmosphere and collide into the surface.

155. There have been 43 unmanned missions to Mars. 21 of them have failed.

156. Mars has seasons.

157. By analysing Mar's recurring slope linae (dark soil streaks running down slopes,) NASA determined that they contain hydrated salts which are formed by water. This water could provide a habitat for microbial life.

158. Pieces of Mars have fallen to Earth as meteorites.

159. Ancient Egyptians called Mars "Her Desher." This means "The Red One."

160. The first space probe to take pictures of Mars' surface was the Mariner 4 in 1964.

161. There is more than enough evidence to prove that Mars used to be covered in water. Probes have located countless dried up lakes and riverbeds and minerals that can only form in water. It probably even had oceans.

So where did it all go?

Well, Mars had a magnetic field just like Earth. The most important word in that statement is *had.* For some reason, Mars' magnetic field shut down. This made the planet vulnerable to solar winds, which eroded the water away over billions of years.

162. So, if Mars had water, did it have life? The robots on the planet have found the minerals necessary to create life but there has been no evidence to verify that anything lived on Mars.

163. The chemical that the robots found which can harbour life are called perchlorate. Ironically, they can harbour microscopic life... but it will make it impossible for human life to sustain itself on Mars. Which means that the only way for human beings to ever live on Mars is not to adapt to it, but to force it to adapt to us.

So how do we do that? Well...

164. The ability to convert a planet so its suits our needs is called terraforming. The literal translation is "Earth-shaping." It would involve altering Mars' atmosphere, soil, temperature, ecology, and topography to match a similar lifestyle to Earth.

165. A flight to Mars would take between seven to nine months.

166. On Mars and Pluto, sunsets are blue.

167. An ex-ventriloquist called Dennis Hope has made millions by selling real estate on the Moon, Mars, and other celestial bodies.

168. The CEO of SpaceX, Elon Musk, intends to send men on a one-way trip to Mars in 2022.

169. In 1952, Wernher von Braun wrote a book called Project Mars where humanity has colonized the red planet. This colonization was led by a man called Elon.

170. Musk wishes to send humans to Mars on a rocket called the BFR. It costs $10 billion to build this vessel. This rocket will also be able to transport passengers anywhere in the world in less than an hour. It could theoretically make a trip from LA to Tokyo in a mere 32 minutes.

171. Musk hopes that a million people will live on Mars by 2060.

172. Mars One is a Dutch non-profit organization that intends to launch four astronauts on a one-way trip to Mars to

establish a permanent human colony there by 2027.

173. Musk theorised that he could nuke Mars' poles to release the water onto the surface in the hopes it would terraform the planet for humanity.

174. If the Mars One mission proved successful, four more colonists would arrive two years later.

175. Mars One claims that they only need $6 billion to accomplish the mission. Not only does this number seem outrageously low, but many funders have pulled out.

176. Mars One have calculated the communication time between Mars and Earth would be at least 20 minutes.

177. Mars One claimed that 200,000 people sent applications to become part of the program. In reality, it was only 2,761.

178. Theoretical physicist, Gerard Hooft, used to be an advisor for the Mars One mission. He quit when he realised that a crewed mission to Mars wouldn't be

possible for a century. On top of that, NASA have absolutely no belief in the Mars One mission.

179. Even if Mars One landed on Mars and built a suit that could withstand the solar radiation of the planet, there's another big problem. We get Vitamin D from the Sun. On Mars, the Sun is so far away, that the astronauts would have a permanent vitamin D deficiency.

180. Since the gravity is lower on Mars, it would create intracranial pressure behind the human eyeball which would eventually make a person blind. So good luck on that mission, Mars One.

Jupiter

181. Jupiter has the shortest day in the Solar System, lasting nine hours and 56 minutes.

182. On average, Jupiter is 484 million miles away from the Sun. At its closest, Jupiter is 460 million miles away from the Sun. At its furthest, Jupiter is 508 million miles from the Sun.

183. At its closest, Jupiter is 346 million miles from Earth.

184. Jupiter weighs two septillion tons. That's 2,000,000,000,000,000,000,000,000 tons.

185. The average temperature of Jupiter's atmosphere is -145 degrees Celsius.

186. A year on Jupiter is 11.9 Earth years.

187. Babylonian astronomers were the first to have records of Jupiter back in 800 BC.

188. The Babylonians called Jupiter "Marduk." Marduk was the central god of

ancient Mesopotamia and he was the god of water.

189. At least nine spacecraft have visited Jupiter.

190. You would be three times heavier on Jupiter.

191. Jupiter moves ten miles per second.

192. In January 2015, the three moons, Io, Europa, and Calisto lined up perfectly in front of Jupiter to form a triple solar eclipse. This won't happen again until 2032.

193. Even though Jupiter is half a billion miles away from the Sun, it is one of the brightest things in the sky. This is because it is so gigantic, that it reflects more sunlight than almost anything in the Solar System.

194. Jupiter's overall mass is 2.5 times greater than every planet in the Solar System.

195. Because of Jupiter's size, it gets hit by debris all the time. A visible impact can be

seen on Jupiter through a telescope almost every year.

196. On September 10th 2010, a video recorded Jupiter being hit by a meteor. The impact is very visible, even from an amateur telescope. It is said to be the most powerful explosion ever recorded in our Solar System.

197. Jupiter was named after the Roman god of thunder. He was the leader of the gods so his name was given to the Solar System's biggest planet.

198. A thousand Earths could fit into Jupiter.

199. Another reason Jupiter deserves its name is because the Roman God, Jupiter, was the protector of the other gods. The planet, Jupiter, protects other planets, especially Earth.

Jupiter is so gigantic that most potential threats to Earth like meteors and comets get caught in Jupiter's gravity. The bigger a meteor is, the more likely it will get ensnared by Jupiter's gravitational pull. Once they become too close to Jupiter, the comets and meteors tend to be ripped apart by Jupiter's gravity or they are hurled into interstellar space. If Jupiter didn't exist, Earth could have been destroyed by an asteroid or meteor many times over.

200. Jupiter's diameter is 88,844 miles. That is 11 times wider than Earth's.

201. Jupiter has 300 times more mass than Earth.

202. Any planet that is similar to Jupiter (a Gas Giant) is called a Jovian planet.

203. Jupiter has countless storms that have been raging for centuries. The most powerful storm is The Great Red Spot. The Spot is usually referred to as a hurricane but it's actually an anti-cyclone.

204. The Great Red Spot changes its size and shape almost every year. Nobody knows why.

205. The Great Red Spot's storm moves at 300 miles per hour.

206. The Great Red Spot is three times larger than Earth.

207. Cyanide makes the Spot look red.

208. The Great Red Spot was spotted three hundred years ago. This means that this storm has been raging for at least three centuries nonstop. For all we know, the storm has been swirling around for millennia.

209. No satellite has ever taken a picture of Jupiter's surface. The pictures you have seen of Jupiter only show the planet's clouds. Technically, Jupiter doesn't have a surface since it's made of gas. If you could pass through the clouds of Jupiter and keep moving down, the planet's gaseous state would become thicker until it took on a liquid texture.

210. The storms on Jupiter are so intense, that the planet looks drastically different ever few months.

211. Jupiter boasts the biggest ocean in the Solar System. The liquid hydrogen ocean makes up 78% of the planet's radius and is 33,884 miles deep.

By comparison, Earth's deepest ocean is the Challenger Deep in the Mariana Trench, which is seven miles deep. That means that Jupiter's deepest ocean is nearly 5,000 times deeper than Earth's.

212. Although astronomers don't know if Jupiter has a core, the planet's centre is hotter than any other planet's, reaching a temperature of about 35,700 degrees Celsius. That's nearly six times hotter than the Sun's surface.

213. Jupiter is made up of bright stripes called Zones and dark stripes called Belts. These stripes swirl in opposite directions. This is very clear if you see a video of Jupiter in motion.

214. Jupiter's gravitational field is so big that the planet is self-sustaining. It doesn't need energy and heat from the Sun like all the other planets. It generates its own energy.

215. It takes 43.2 minutes for the Sun's light to reach Jupiter.

216. Some scientists believe that Jupiter is so big, that it nearly became a star. Some astronomers considered it "a failed star."

Although Jupiter is gigantic, it is far too small to be a star.

217. Weirdly, Jupiter emits more heat than it receives from the Sun. It is the only planet to do this in the Solar System. Every other planet gives and takes the same amount of heat.

218. As I write this, Jupiter currently has 79 confirmed moons. When I was a kid, it was 32. By the time you read this, it will probably be a hundred. Jupiter's moons are called Metis, Adrastea, Amalthea, Thebe, Io, Europa, Ganymede, Callisto, Themisto, Leda, Himalia, Lysithea, Elara, Dia, Europie, Thelxinoe, Euanthe, Helike, Orthosie, Iocaste, Praxidike, Harpalyke, Mneme, Carpo, Hermippe, Thyone, Ananke, Herse, Aitne, Kale, Taygete, Chaldene, Erinome, Aoede, Kallichore, Kalyke, Carme, Callirrhoe, Eurydome, Pasithee, Kore, Cyllene, Eukelade, Pasipha, Hegemone, Arche, Isonoe, Sinope, Spone, Autonoe, and Megaclite. The other moons have yet to be named.

219. In 1610, Galileo saw three objects circling around Jupiter through his

telescope. A week later, he discovered a fourth object. This was a HUGE deal because it proved that not everything in the universe revolved around the Earth… literally. These objects were Jupiter's moons, Ganymede, Callisto, Io, and Europa. They became known as the Galilean Moons.

220. Callisto is the most cratered object in the Solar System. It is the moon version of a pimply teenager.

221. Ganymede is the biggest moon in the Solar System, measuring 3,274 miles across. It's bigger than Mercury and almost as big as Mars. It's so big, it's the only moon that has a magnetic field.

222. A salted ocean lies beneath the surface of Ganymede. Images from the Hubble telescope show that certain features of the moon's surface were probably formed by flood water emerging through faults or cryo-volcanoes. (A cryo-volcano spews ice instead of lava.) The European Space Agency has announced that funding is available for the Jupiter Icy Moons Explorer (which is known as JUICY) to launch probes to explore the body and other moons by 2030.

223. Io orbits Jupiter so tightly, that it can revolve around the planet in 42 hours (which is astounding considering Jupiter's vast size.)

224. Jupiter has Northern and Southern Lights like Earth.

225. Although it is not visible in pictures, Jupiter has four rings.

226. Io has magma oceans and 400 active volcanoes. It is the most volcanic moon or planet in the Solar System. The moon is so volcanic, at least one volcano is erupting at any one time. Some of Io's volcanic blasts are so powerful, that they can spew lava out of the atmosphere. These eruptions can be visible from space. Lava spews so regularly on Io, that the moon's surface changes dramatically every few months.

227. Since Io is basically a planet-sized volcano, you would assume that it would sound like an eruption?

Nope. Sounds have been picked up from Io's atmosphere. Do you know what it sounds like? A truck reversing. Seriously.

228. You might wonder how Io has so much energy. Why doesn't it run out of lava? Earth has many dead volcanoes. How can Io's volcanoes still be active? Oddly, it's because of Jupiter and the other moons. Since so many moons hurtle past Io, they pull it one way while Jupiter pulls it the other way. This stretches and squashes Io which heats it up through friction.

229. Jupiter's moon, Europa, has more water under its surface than all of Earth's oceans.

230. Europa's water is salty. Since life on Earth began in salt water, it's possible that the waters of Europa harbour life. The life would probably be basic like micro-organisms or, at best, fish, but any form of life outside of Earth is fascinating to astronomers. To properly explore Europa, scientists want to build a squid-like robot to explore its oceans.

231. Jupiter's atmosphere sounds like someone blowing out of a didgeridoo.

232. Hundreds of asteroids orbit Jupiter. They are too small to be classified as moons.

233. The moon, Amalthea, is so close to Jupiter that the planet would cover half the sky if you stood on Amalthea's surface.

234. Some of Jupiter's moons are so tiny, it's been theorized that they used to be asteroids that escaped from the Asteroid Belt when they got snagged by Jupiter's gravity. This makes sense as tiny moons like Thebe, Kalichore, Kale, Hegemone, and Pasipha are irregularly shaped unlike the other spherical moons. Another theory suggests that these tiny moons use to be fused together before they were struck by a meteor or comet.

Saturn

235. Saturn is named after the Roman god of agriculture. His Greek name is Cronus.

236. On average, Saturn is 889 million miles from the Sun. At its closest, Saturn is 839 million miles from the Sun. At its furthest, Saturn is 938 million miles from the Sun.

237. At its closest, Saturn is 746 million miles from Earth.

238. It takes 1 hour 19 minutes and 3 seconds for the Sun's light to reach Saturn.

239. A day on Saturn is 10 hours and 39 minutes.

240. It takes Saturn 29.45 years to orbit the Sun.

241. Saturn's diameter is 75,000 miles.

242. 755 Earth's could fit into Saturn.

243. Saturn's core is 11,700 degrees Celsius.

244. Saturn moves eight miles per second.

245. Saturn is twice as far away from the Sun as Jupiter.

246. Saturn's average temperature is -178 degrees Celsius.

247. Saturn bulges at the equator (much like Earth.) However, Saturn bulges a full 10% more at the equator than at the poles. It's very noticeable if you looked at the planet through a telescope.

248. There is a hurricane on Saturn that has been raging since at least 2013. The hurricane is 12,500 miles across. That's 20 times larger than any recorded hurricane on Earth. Even the eye of the hurricane is 1,250 miles across. The wind in this hurricane moves at 300 miles per hour.

249. Earth's hurricanes are powered by warm oceans and dissipate over cold waters. Saturn has no oceans. As a result, scientists don't have a clue how Saturn's hurricane can exist.

250. In 2010, a storm was discovered in the northern hemisphere of Saturn. The storm

was so big, it literally wrapped around the entire planet, measuring nearly 200,000 miles in length.

251. Saturn was observed by Assyrians in 800 BC.

252. The Babylonians believed Saturn was a star and worshipped it.

253. The poles are hexagonal in shape and are known as Mega Hexes.

254. Galileo discovered Saturn's rings. Unfortunately, his telescope wasn't advanced enough for him to figure out what they were. He thought Saturn was three planets merged together.

255. Dutch physicist, Christiaan Huygens, deduced that Saturn had rings in 1655. He also discovered its moon, Titan.

And just to show off, he also pioneered the use of pendulums in clocks.

256. Although it's the second biggest planet in the Solar System, Saturn is the least dense. It's so light, it would float on water.

257. Saturn's rings are made of countless clumps of ice. Each ice rock that makes up Saturn's rings only measures about ten metres.

258. Saturn's rings measure 155,000 miles across. That's 2/3rds the distance between the Earth and the Moon.

259. It is unclear why Saturn has rings. Most astronomers believe that Saturn's rings formed when a huge moon in the planet's orbit was destroyed by an asteroid or comet.

260. Saturn has three main rings. They are called A, B, and C. (Astronomers are surprisingly uncreative with names.) B is the biggest ring, measuring 1,600 miles wide. There is a 3,100-mile gap between B and C. This gap is known as the Cassini Division.

261. Millions of years in the future, Saturn will absorb its own rings.

262. Saturn's atmosphere makes the WEIRDEST sound. It sounds like an echoing whistle. Sometimes, the planet sounds like someone adjusting an old-fashioned radio signal machine. If you heard the sound out of

context, you would assume it was from a cheesy sci-fi movie from the 1950s.

263. Usain Bolt's speed would allow him to fly on Saturn's moon, Titan.

264. Saturn has 62 moons. 53 are named. Most of them are named after Norse gods. The 53 named moons of Saturn are Pan, Daphnis, Atlas, Prometheus, Pandora, Epimetheus, Janus, Aegeon, Mimas, Methone, Anthe, Pallene, Enceladus, Tethys, Telesto, Calypso, Dione, Helen, Polydeuces, Rhea, Titan, Hyperion, Iapetus, Kiviuq, Ijiraq, Phoebe, Paaliaq, Skathi, Albiorix, Bebhionn, Erriapus, Skoll, Siarnaq, Tarqeq, Greip, Hyrrokkin, Jarnsaxa, Tarvos, Mundilfari, Bergelmir, Narvi, Suttungr, Hati, Farbauti, Thrymr, Aegir, Bestla, Fenrir, Surtur, Kari, Ymir, Loge, and Fornjot.

265. The moon, Iapetus, is shaped like a walnut.

266. Hyperion looks like a block of Styrofoam.

267. Hyperion was discovered by William Cranch Bond. He was one of the earliest

American astronomers and discovered this moon despite being born into poverty and having no formal education. He eventually became the first director of the Harvard College of Observatory.

268. The moon, Mimas, looks suspiciously similar to the Death Star from Star Wars.

269. The Cassini spacecraft has been orbiting Saturn for over 11 years. This is the longest time that any spacecraft has surveyed an outer planet. It ran out of power in 2017.

270. Saturn's biggest moon is Titan, measuring 3,200 miles across. It is also the only moon to have an atmosphere.

271. Although Titan was discovered in 1655, it didn't have a name until 1847.

272. When the Huygens spacecraft landed on Titan's surface, the team controlling it learned that the moon has an underground ocean of liquid water. Dr. Ralph Lorenz at the Lunar and Planetary Science Convention wants to build a submarine to explore Titan's waters. NASA are strongly considering this idea.

273. Enceladus reflects all the Sun's light, making it the shiniest moon in the Solar System.

274. Enceladus has geysers that spew ice into the atmosphere. These geysers developed in chasms and are known as Tiger Stripes. The geysers of Enceladus are so powerful that they form one of Saturn's rings. The ice from the geyser makes what is known as the E-ring.

275. Chiron is an alleged moon that Hermann Goldschmidt claimed to see in 1861 but it hasn't been seen since.

276. Themis was supposedly sighted in 1905 by astronomer, William Pickering.

No one else has ever seen it and it is unlikely Themis or Chiron exist.

277. The most common reason astronomers give to explain why they became fascinated by space is, "I saw Saturn through a telescope."

Uranus

278. Uranus is named after the Greek god of the sky. His Roman name is Caelus.

279. On average, Uranus is 1.79 billion miles from the Sun. At its closest, Uranus is 1.71 billion miles from the Sun. At its furthest, Uranus is just under 1.86 billion miles from the Sun.

280. At its closest, Uranus is 1.6 billion miles from Earth.

281. Uranus's average temperature is -224 degrees Celsius.

282. Uranus' core is 4,737 degrees Celsius.

283. Uranus takes 84 years to travel around the Sun.

284. It takes 2 hours 39 minutes and 6 seconds for the Sun's light to reach Uranus.

285. Uranus's name has been the butt of many jokes (no pun intended.) Weirdly, because of the heavy level of methane in

Uranus' atmosphere, the planet smells like rotten eggs.

286. Most people mispronounce Uranus. It's pronounced "YUUR-a-nis."

287. Uranus' diameter is 31,763 miles.

288. The weirdest thing about Uranus is that it's sideways. No one knows why. The most obvious theory is it got hit by another planet which would've knocked it off its centre. However, Uranus isn't very solid so an impact shouldn't have had that effect. Unfortunately, there is no other logical theory.

289. Uranus has a ring system like Saturn. The ring isn't horizontal like Saturn's but vertical (because the planet rotates vertically.) The rings were discovered in 1977.

290. Uranus has 13 rings. They are mostly made of ice. They were probably created when one of Uranus' old moons was destroyed by a comet or asteroid.

291. Uranus moves six miles per second.

292. One day on Uranus lasts for 17 hours and 14 minutes.

293. Uranus' original name was George's Star. It was named after King George III.

294. Uranus is classified as an Ice Giant.

295. It rains diamonds on Uranus. No, really.

296. Between August 6th 2014 – August 6th 2015, storms appeared on Uranus that were so powerful, that they were visible through telescopes on Earth.

297. If you listen to the radio waves from Uranus, it sounds like rustling trees.

298. Uranus has 27 moons. They are named after characters created by William Shakespeare and Alexander Pope. The moons are called Cordelia, Ophelia, Bianca, Cressida, Desdemona, Juliet, Portia, Rosalind, Cupid, Belinda, Perdita, Puck, Mab, Miranda, Ariel, Umbriel, Titania, Oberon, Francisco, Caliban, Stephano, Trinculo, Sycorax, Margaret, Prospero, Setebos, and Ferdinand.

299. The moon, Miranda, has the tallest cliff in the Solar System, measuring five miles high. That would take a human up to six minutes to fall to the bottom.

300. Miranda is known as the Ugly Duckling of the moons. Miranda looks like a bunch of rocks stuck together with superglue.

301. Miranda's interaction with Uranus' magnetosphere makes a constant reverberating echo sound.

302. Based on gravitational waves interfering with Uranus' rings, planetary scientists, Rob Chancia and Matthew Hedman, theorise that there are two other moons in Uranus' orbit that we have yet to discover.

303. Even though Uranus is the second-farthest planet from the Sun, it was the first planet to be discovered. How is this possible? Well, even though Mars and Venus have been visible from Earth for millennia, they were regarded as unusual stars, not planets. Uranus was discovered accidentally by William Herschel in 1781 while he was studying stars. He noticed that Uranus

moved like a planet and it was too small to be a star. Bizarrely, Uranus, was discovered before Antarctica.

Neptune

304. Neptune is named after the Roman god of the sea. His Greek name is Poseidon.

305. On average, Neptune is 2.8 billion miles from the Sun. At its closest, Neptune is 2.77 billion miles from the Sun. At its furthest, Neptune is 2.83 billion miles from the Sun.

306. At its closest, Neptune is 2.7 billion miles from Earth.

307. It takes four hours for the Sun's light to reach Neptune.

308. Neptune's temperature is -218 degrees Celsius.

309. Neptune's core is 7,000 degrees Celsius.

310. Neptune's diameter is 30,775 miles.

311. The wind moves at 1,300 miles per second on Neptune. That's about the same speed of sound.

312. Neptune has a ring system like Saturn. However, it is not visible in photographs.

313. Of all the planets in the Solar System, Neptune and Uranus share the most similarities. They are even in the same classification – Ice Giants.

314. Neptune is slightly smaller than Uranus. However, Neptune is slightly denser than Uranus.

315. If you could stand on Neptune's surface, the Sun would be visible but it would be almost indistinguishable from other stars.

316. One day on Neptune lasts for 16 hours and 6 minutes.

317. Galileo saw Neptune in 1612. This means they he observed the planet 234 years before it was officially discovered.

318. It takes Neptune 165 years to go around the Sun. Since its discovery in 1846, Neptune has only gone around the Sun once.

319. Neptune moves three miles per second.

320. Neptune is 14.5 times bigger than Earth.

321. Neptune is the only planet that is completely invisible to the naked eye from Earth.

322. Neptune was discovered in 1846 by a French mathematician called Urbain le Verrier. Weirdly, he didn't discover Neptune by accident. He realised that Uranus's position in the Solar System didn't seem mathematically correct so he believed that a large planet was interfering with its location. Neptune is the only planet in the Solar System to be discovered with math.

323. John Couch Adams also used maths to locate Neptune's position but when he revealed his calculations, he learned that Neptune was already discovered... two days before.

324. Many sounds have been heard from Neptune. Sometimes it sounds like intense blizzard, a hurricane, or a whistling noise. If you listen to Neptune on YouTube, you can even hear some bleeping sounds.

325. Neptune has 14 moons. One of them is unnamed. Neptune's named moons are called Naiad, Thalassa, Despina, Galatea,

Larissa, Proteus, Triton, Nereid, Halimede, Sao, Laomedeia, Psamathe, and Neso.

326. Triton is Neptune's biggest moon. It's 1,677 miles across.

327. Triton is -235 degrees Celsius. This is the coldest natural temperature in the Solar System.

328. For some reason, Triton orbits around Neptune backwards. None of Neptune's other moons do this.

329. Most of what is known about Triton came from the single visit of Voyager 2 in 1989. Voyager 2 took pictures of 40% of the moon's surface.

330. Triton is gradually getting closer to Neptune. Eventually, it will become too close and be torn to shreds and form a ring around the planet more spectacular than Saturn's.

331. The last moon to be discovered was the 14th moon of Neptune on July 15th 2013. It doesn't have an official name yet.

Pluto

332. Pluto is named after the Greek god of the underworld. The Ancient Romans called him Hades.

333. On average, Pluto is 3.67 billion miles away from the Sun. At its closest, Pluto is 2.76 billion miles from the Sun. At its furthest, Pluto is 4.59 billion miles from the Sun.

334. At its closest, Pluto is 2.66 billion miles from Earth.

335. It takes 5 hours and 30 minutes for the Sun's light to reach Pluto.

336. Pluto was discovered by Clyde William Tombaugh in 1930.

337. Until Pluto was found, it was theorised it would be a gigantic planet like Uranus or Neptune. That discovery must've been pretty anti-climactic.

338. Pluto is -230 degrees Celsius on the surface.

339. Pluto was discontinued as a planet on August 24th 2006. There are three rules in order for a celestial body to be considered a planet -

i) A planet must be in orbit around the Sun.

ii) The object must be massive enough to be rounded by its own gravity.

iii) It must have cleared the neighbourhood around its orbit.

Pluto fails the third condition.

340. Pluto moves two miles per second.

341. Pluto is now regarded as a dwarf planet. The are other dwarf planets beyond Pluto called Makemake, Haumea, Eris, and Ceres. Pluto is the largest dwarf planet.

342. At least seven moons in the Solar System are bigger than Pluto. It is half the size of our Moon.

343. Pluto has five moons; Charon, Styx, Kerberos, Nix, and Hydra.

344. There are floating hills on the surface of Pluto. These are icebergs on a sea of frozen nitrogen.

345. Pluto will complete just one orbit of the Sun since its discovery in 2117.

346. It takes Pluto 248 years to travel around the Sun.

347. Pluto has a crater on its surface in the shape of a love-heart.

348. A day on Pluto is six Earth days and nine hours long.

349. Pluto is 1/3rd water.

350. Pluto is the only planet to have been discovered in the 20th century even though it isn't a planet anymore.

351. Pluto was named by an 11-year-old girl from Oxford called Venetia Burney.

352. Since Pluto lost its status as a planet, it has also lost its name. It's now officially called 134340.

353. On April Fool's Day, 1976, BBC announced that Pluto would pass behind Jupiter, which would reduce the gravity of

Earth. This would cause people to float if they jumped at exactly 947am. Although it was an obvious joke, hundreds of people called the BBC, claiming that they hovered for a few seconds due to the "Jovian-Plutonian gravitational effect."

The Moon

354. For years, the most popular theory on how the Moon formed was is the Giant Impact Hypothesis. The theory states that 30 million years after Earth formed, a Mars-sized planet called Theia grazed our world. Debris from Theia and Earth melded together to form the Moon.

In recent years, astronomers have begun to doubt this theory and are trying to come up with another hypothesis.

355. The Moon is the only surface apart from Earth's that has been stepped on by humans. The footprints on the Moon will last for millions of years.

356. The Moon is moving away from the Earth by four centimetres a year.

357. In 300 BC, Aristotle concluded that the Moon affects the tides.

358. The Moon is 235,000 miles from Earth.

359. The side of the Moon that we can't see is turquoise-colored.

360. The Moon has no atmosphere.

361. Even though there are hundreds of moons in the Solar System, we call ours The Moon… which is pretty rude to the other moons. The Moon's actual name is Luna.

362. Although the Moon is the fifth largest moon in the Solar System, it is the biggest moon in size compared to its orbiting planet. Although Jupiter's moon, Ganymede, is bigger than our Moon, it's tiny compared to the planet itself.

363. There is a misconception that a full Moon makes some people and animals go wild. (The word "lunatic" is derived from the Moon's name, Luna.) This idea comes from the philosopher, Aristotle. Since Aristotle knew the Moon affected the tides, he believed the Moon was affecting the water in the brains of all living things. But Aristotle thought the womb moved around the body and snot was liquefied brain matter so he is not a particularly reliable source of information. 37 separate studies have been done that prove that the Moon has no effect on people's mental stability.

But wait... if the Moon has no effect on animals, why do they go crazy sometimes during a full Moon? It is true that wild animals tend to go berserk after a full Moon but not for the reason you might think. The fuller a Moon is, the brighter it is. The brighter the Moon is, the easier it is for animals to see. If animals can see at night, it's more likely that they can escape from predators. So if you hear reports that animals have gone wild after a full Moon, it's not because the Moon drove them mad; it's because they are starving!

364. The Moon is about the same size as Australia.

365. You would weigh six times less on the Moon (16.5%.) A 180lb person would weigh 30lbs on the Moon.

366. Only 12 people have walked on the Moon. Every person who has walked on the Moon was an American man.

367. The Moon is 2,323 miles in diameter. That's about a quarter the size of Earth's diameter.

368. 7% of Americans think the Moon landings were faked.

369. The Moon suffers tremors called moonquakes from the gravitational pull of the Earth.

370. Many reports state that the last thing said on the Moon was, “God willing, we shall return.” In reality, the last thing said on the Moon was, “All right, let’s get this mother out of here and go home.”

371. In 1958, the Air Force hatched Project A119; A Study of Lunar Research Flights. Basically, they wanted to nuke the Moon.

During the height of the Cold War, the US wanted to show the power of their military arsenal by detonating a nuclear weapon on the Moon. This detonation would have been visible from Earth. After eight project reports were commissioned, they concluded that it could be done… but they decided not to… because that would be insane.

372. The Moon is divided into two sections – the Highlands and the Maria. The Highlands

are heavily cratered from asteroid impacts and the Maria is smooth and dark.

373. During the Apollo 15 and 17 missions in 1971-1972, the Moon's surface temperature increased. For decades, nobody knew why. The mystery was only solved in 2010 by Seiichi Nagihara. After analysing data for years, he concluded that human interference was to blame. Landing on the Moon disturbed the Moon's surface, which exposed the soil underneath. This soil absorbed heat from the Sun for the first time in millions of years, causing the Moon's surface temperature to increase by one or two degrees Celsius.

374. We always see one side of the Moon. It is known as the Daylight Side or the Bright Side. The side of the Moon that we cannot see is known as the Dark Side of the Moon. The line that divides the Bright Side and the Dark Side is called the Terminator. The Dark Side of the Moon was seen for the first time in 1959 by the Soviet's Lunar 3 space probe. The Dark Side looks very different from the side of the Moon that we see. Not only does it have barely any Maria but the crust itself is much thicker.

375. The Moon isn't spherical. It's shaped like an egg. It appears spherical because we always see the same side of it.

376. The Moon has a volcano on the South Pole that died millions of years ago.

377. The word "month" comes from the word "moon." In most languages, the words "month" and "moon" sound similar.

378. Everyone's heard of a rainbow but did you know that moonbows exist? When droplets of water combine with mists made from waterfalls, a night rainbow can be visible during a full Moon.

379. When the Moon is on the horizon, it can look twice the size as when it is high up. This is called the Moon Illusion because it is actually the same distance from Earth whether it's high or low.

380. Galileo was the first astronomer to notice that the Moon had craters. This annoyed people as they believed the Moon was a perfect sphere created by God.

381. Most of the craters were created during the early development of the Solar System. Many asteroids and comets smashed into the Moon's surface 4.1 billion years ago. This was known as the Heavy Bombardment.

382. The Moon's core is about 218 miles in width.

383. NASA beamed Wi-Fi Internet to the Moon, just to see if they could. Its speed was 19MB per second.

384. Despite what many trivia books say, you can't see the Great Wall of China from the Moon. You can barely see the Great Wall from Earth's orbit. Also, 99% of the Great Wall is crumbled to dust so you can barely see it even when you're standing on it.

385. The Moon moves around the Earth at the speed of a rifle bullet.

386. The oldest crater on the Moon is over four billion years old.

387. The biggest crater is Tycho, measuring 932 miles across. That's nearly half the size

of the Moon. It's so big, it can be seen during a full Moon.

388. Sometimes, the Moon turns a blood-red colour. This is because the Earth is starting to block the sunlight heading towards the Moon and the only light that can get through comes through the thickest part of our atmosphere.

389. There seems to be empty river channels on the Moon's surface. Astronomers believed that these channels may have had water. It was later discovered that they harboured lava from the Moon's core. When asteroids collided with the Moon, lava bubbled up from the core and seeped out onto the surface.

Any water on the Moon in the past evaporated away over time. However, some craters can be so deep that they can collect water in the form of ice. The deepest craters on the Moon are at the poles and may contain up to a billion tons of water.

390. A solar eclipse is when the Moon is in front of the Sun from the viewpoint of the Earth. A lunar eclipse is when the Sun is in

front of the Earth from the viewpoint of the Moon.

391. The Sun is approximately 400 times wider than the Moon. By a complete coincidence, the Sun is also 400 times farther away from the Earth than the Moon. This is why the Sun and the Moon appear to be the same size in the sky. This incredible coincidence is how solar eclipses are possible. Because the Moon is the same size as the Sun in our sky, the Moon can completely cover the Sun.

392. It takes more computing power to perform a Google search than to send Apollo 11 to the Moon.

393. Solar eclipses occur about twice a year but can happen up to five times a year.

394. A solar eclipse only lasts for seven minutes at most.

395. When the solar eclipse is about to end and the Sun starts to appear again, it looks like a glowing wedding ring. This is known as the Diamond Ring Effect.

396. When the Moon appears slightly smaller than the Sun during an eclipse, the edges of the Sun are just visible, making the eclipse look like a ring. This is called an annular eclipse.

397. Looking at an eclipse directly can be dangerous. Obviously, looking at the Sun can hurt your eyes. But what's worse than that is when you look at an eclipse in almost complete darkness and then suddenly, the bright Sun appears again. It's like when you turn on a very bright light in a pitch-dark room and it distorts your vision (except it's much worse with the Sun because it's a nuclear explosion.) Looking at the eclipse during this moment won't make you go blind as some sources suggest but it can damage your retinas.

398. All astronauts say the Moon smells like gunpowder.

399. Some people may ask, "What is the point of going to the Moon?"

The short answer is Helium 3 (also known as He3.) He3 is a lightweight isotope that could single-handedly fix energy problems on Earth. NASA is trying to find a way to mine it successfully on the Moon.

400. When the right side of the Moon is crescent-shaped, astronomers call this a Waxing Crescent.

401. When you can only see half of the Moon's front, most people call it a half-Moon. However, since we only see one side of the Moon, technically you are only seeing 1/4th of the Moon. When only the right side of the Moon is visible, astronomers call it First Quarter.

402. When most of the right side of the Moon is visible, it is known as Waxing Gibbous.

403. When most of the left side of the Moon is visible, it is known as Waning Gibbous.

404. When only the left side of the Moon can be seen, astronomers call it Third Quarter.

405. When the left side of the Moon is crescent-shaped, astronomers call it a Waning Crescent.

406. When the Sun, Earth, and Moon are aligned, it is known as a New Moon.

407. If you were standing on the Moon and looked at a full Earth, it would be 50 times brighter than a full Moon. This is called Earthshine.

408. The next mission to the Moon will be in 2020. Luna 27 will land on the South Pole of the Moon to collect frozen water.

409. Luna 9 was the first spaceship to land on the Moon. Other spacecraft had crashed into the Moon before that.

410. The fastest speed recorded on the Moon is 10.5 miles per hour by the Lunar Rover Vehicle.

411. The Ancient Greeks called the Moon "Selene."

412. People who are scared of the Moon suffer selenophobia.

413. In 2017, Elon Musk began discussions on building a base on the Moon.

414. 34-year-old Chilean lawyer, Jenaro Gajardo Vera, went to his registration agency in 1953 and bought the Moon for the equivalent of $2. Since nobody had ever bought the Moon legally, no one could challenge Vera and so, the celestial body became his. Vera bought the Moon so he could get into a nightclub that only accepted members who owned property.

415. The cosmetics company, Foreo Institute, want to make the Moon brighter, so humanity can save money on power.

416. Neil Armstrong walked on the Moon on July 21st 1969. You may have heard of him.

417. While orbiting the Moon, the commander, James A. Lovell Jr. accidentally used up some of the memory on the on-board computer, which caused the thrusters to go haywire. Lovell had no choice but to land the ship on the Moon manually using the position of the stars to align the craft.

418. While on the Moon, the shadows are so dark, astronauts couldn't see their hands in front of them.

419. Putting a man on the Moon cost $25.4 billion.

420. NASA scientists said that the most dangerous thing about the Moon was the dust. The dust was strong enough to tear through three layers of bullet-proof material on astronauts' boots and could clog up joints on space suits. Moon dust is toxic to the lungs, making it the biggest dilemma in terms of long-term stays on the Moon.

421. Neil Armstrong was meant to say, "That's one small step for *a* man, one giant leap for mankind." Originally, Armstrong said the static obscured the "a" when he said this famous line but after hearing it, he realized that he forgot to say the "a."

422. Neil Armstrong's application to NASA was delayed by a week. He was only accepted because his friend, Dick Day, secretly inserted Armstrong's application into the pile.

423. The Apollo 11 footage of Neil Armstrong and Buzz Aldrin landing on the Moon lasted 153 minutes.

424. Neil Armstrong walked on the Moon for 19 minutes before Buzz Aldrin took a step on it.

425. Neil Armstrong placed an American flag on the Moon. However, it was too near Armstrong's lunar module when the ship lifted off and so, the flag was blown away.

426. There is a famous picture of a boot imprint on the Moon. This imprint was made by Buzz Aldrin.

427. The flag that was erected on the Moon cost $5.50.

428. In 2006, NASA admitted that they no longer have the original tapes of the Moon landing because they taped over them.

429. When Neil Armstrong, Buzz Aldrin, and Michael Collins returned from space, they were quarantined for three weeks in case they were exposed to microbes.

430. Armstrong, Aldrin, and Collins were in space for 8 days, 14 hours, 12 minutes and 30 seconds.

431. Weirdly, Buzz Aldrin's mother's maiden name was Moon.

432. The most famous picture of an astronaut on the Moon is incorrectly believed to be of Neil Armstrong. In fact, this iconic image is of Buzz Aldrin. Armstrong took the picture and he can be seen in the reflection of Aldrin's helmet visor.

433. Astronauts aboard Apollo 10 said they heard music while orbiting the Dark Side of the Moon. It turned out to be radio interference between the Lunar Module and the Orbiter.

434. 88% of the astronauts who have walked on the Moon were Boy Scouts.

435. If the Moon didn't exist, days on Earth would only last six hours.

436. The Moon is 18 times further away from the Earth than when it first formed 4.5 billion years ago.

437. The Apollo 11 astronauts have a moon on the Hollywood Walk of Fame instead of a star.

438. Star Trek fans complained that the show was interrupted by footage of Neil Armstrong walking on the moon. There is so much irony in that fact.

439. The Apollo 11 astronauts went to the Moon without life insurance. Instead, they signed hundreds of autographs and gave them to family members to sell in case they died.

440. Church of Mormon founder, Joseph Smith, said that there were men dressed as Quakers living on the Moon. This is probably not true.

441. In 1961, another Mormon (also called Joseph Smith) said that no man will ever reach the Moon. NASA strongly disagree.

442. In 1835, the New York Sun newspaper reported that a new telescope showed that the Moon was populated with winged-men and unicorns.

Astronomers

443. Although Galileo tends to get the credit for inventing the telescope, it was concocted by Hans Lippershey in 1608.

Galileo heard of the invention and made a much better version several years later.

444. The most famous telescope is the Hubble Telescope. It was named after American astronomer, Edward Powell Hubble, who proved that there were galaxies beyond the Milky Way. His greatest discovery was the relationship between a galaxy's distance from Earth and the speed that it is moving.

445. The first record of the constellations was in Poetica Astronomica. It was written by Gaius Julius Hyginus between 64 BC-17 AD. It was the first document to have the lore about the constellations (Taurus, Gemini, Scorpio, etc.)

446. Kip Thorne is a theoretical physicist and one of the greatest astronomers of today. He was even a producer on the movie, Interstellar, to keep it as realistic as possible.

447. On September 5th 1862, James Glaisher tried to go to space in a hot-air balloon. He didn't succeed.

448. William Herschel is most famous for discovering Uranus but he also discovered infrared light, which is incredibly useful in finding celestial bodies in the universe. He also discovered 800 stars and 2,500 nebulae.

449. William Herschel's sister, Caroline, worked closely with him. Although she started off as Herschel's assistant, she became an important astronomer and discovered several comets, was awarded the Gold Medal of the Royal Astronomical Society, and named an Honorary Member of that Society and the Royal Irish Academy.

NASA

450. NASA is an acronym for National Aeronautics and Space Administration. The term "aeronautic" originated from France, and was derived from the Greek words for "air" and "to sail."

NASA used to be called NACA, which stood for National Advisory Committee for Aeronautics.

451. NASA was opened by President Dwight D. Eisenhower on October 1st, 1958. He saw it as an educational project rather than a military one.

452. NASA originally had four laboratories and 80 employers.

453. NASA's headquarters are in Washington DC.

454. Currently, NASA has bases for Research Centre, Test Facilities, Construction and Launch Facilities, and Deep Space Network stations.

455. Space shuttles were developed and launched in Florida.

456. NASA's very first space shuttle was Columbia in 1981.

457. NASA's Discovery has flown more space missions than any other spacecraft. Discovery's first journey was in 1984 and its 39th and final journey was in 2011.

Discovery was in space for a total of 365 days and it travelled 148,221,675 miles. That's 600 times longer than the distance between the Earth and the Moon.

458. NASA will pay volunteers $15,000 to lie in a bed for 24 hours a day for 90 days to measure the effects zero gravity has on a person's body.

459. NASA hired 1,600 Nazi scientists to build supersonic rockets, guided missiles, and nerve gas.

460. A habitable artificial satellite called the International Space Station was launched into space in 1998. It cost $150 billion, making it the most expensive object ever made. It's 239ft long, 356ft wide, and 66ft tall. It was built by NASA, Roscosmos, JAXA, ESA, and CSA.

461. The International Space Station travels at five miles per second.

462. Spaceship cargo costs $20,000 per pound.

463. The toilets in a NASA spaceship cost at least $50,000 each.

464. NASA studied giraffes because their leg veins grow incredibly quickly. Understanding how giraffes' legs work allowed NASA to create Lower Body Negative Pressure. This concept allowed astronauts to float in space without suffering from blood flow problems. Thanks, giraffes.

465. As a management training program, NASA trainees watch the movie, Armageddon, to try and find as many scientific inaccuracies as possible. There are 168 of them.

466. One of the founders of NASA, Jack Parsons, was an occultist who recited pagan hymns to the nature God, Pan, during his

rocket trials at work. He was friends with the creator of Scientology, L. Ron Hubbard.

467. Polls showed that Americans believed that NASA's funding accounted for 20% of the federal budget. It is less than 1% and has been since 1975.

468. NASA sent spiders into space to see how long it would take them to learn how to weave webs in zero-gravity. It only took the spiders two days.

469. The Internet at NASA has a speed of 91 GB per second.

470. NASA uses thermal tiles to prevent spaceships from burning up when they re-enter Earth's atmosphere. Although the inside of these tiles can get up to 1,200 degrees Celsius, their insulating properties are so good that it's perfectly safe to hold one. It is the hottest object a human being can hold without being harmed.

471. Because NASA studies radiation and how it affects the human body, NASA has helped the fight against breast cancer.

472. NASA developed a 3D printer that makes pizzas for astronauts.

473. A spacesuit's purpose is to ensure an astronaut can survive the vacuum of space. So how did NASA test this before they ever sent people to space?

Simple. They got a volunteer called Jim LeBlanc to wear a spacesuit and walk into a sealed vacuum chamber. The test worked... for a few seconds. When LeBlanc's hose disconnected, he felt "the saliva on my tongue starting to bubble" due to the change in atmosphere pressure. Luckily, he survived. He is one of the only people to ever survive this level of atmosphere change.

474. NASA have invented many things including modern sunglasses, cordless tools, purified water, superior baby food, heavily insulated firefighter suits, frozen dried food, pacemakers, blood analyzers, advanced wheelchairs, prosthetic limbs, improved hearing aids, virtual reality helmets, shoe insoles, and camera phones.

475. In 1962, a programmer forgot to type a hyphen on the code for the Mariner I rocket.

This caused the rocket to explode as it took off, costing NASA $630 million.

476. Your smartphone is more powerful than all the technology that NASA had when they sent men to the Moon.

477. The first animal that NASA sent into space was a fruit fly. In 1946, fruit flies were sent into space to test radioactive exposure to see if humans could survive space.

478. Laika the dog was sent into space in 1957.

479. When NASA's spaceships re-enter Earth's atmosphere, the ships burn up. It's not the atmosphere that creates the heat, but the compression. As the craft hurtles through the sky, the air around it compresses and heats up so quickly, it doesn't have a chance to cool until it slows down.

480. When the mirror for the Hubble telescope was being made, a microscopic chip of paint flicked off a measuring rod in a device called a "null corrector." As a result, the mirror was slightly too flat – but only by

four microns (50 times thinner than a human hair) at the outer edge. Although a micron is tiny, space is MASSIVE, so the images it made were very blurry. NASA had to spend millions of dollars to fix the telescope.

481. The first Apollo mission was a failure. The three astronauts testing Apollo 1 died in a fire in the capsule.

482. Apollo 4-6 were unmanned missions.

483. Apollo 7 was a manned mission but it didn't go to the Moon.

484. Apollo 8-10 went to the Moon but they were not designed to land.

485. Apollo 11-17 were manned missions sent to land on the Moon.

486. The Apollo 11 lunar module had 20 seconds of fuel left when it landed.

487. NASA hired a man called George Aldrich to sniff everything they send into space. If he doesn't like the smell, it doesn't get sent into space.

488. NASA commanders are required to have logged 1,000 jet aircraft flight hours before they are considered ready to fly a spacecraft.

489. There is an urban legend stating that ballpoint pens don't work in zero gravity so NASA spent $1.5 million to perfect a Space Pen. Russians just used pencils.

Although this is a funny story, it's not true. Although a Space Pen was invented for over a million dollars. NASA didn't pay a penny. Paul C. Fisher of Fisher Pen Co. did this of his own free will. He was offering his services and wasn't asked by NASA for help. He created the Space Pen and gave it to NASA.

Astronauts can't use pencils because the tip can flake off. The smallest form of debris in a space capsule can be hazardous because it could float into an astronaut's eye or into electric equipment.

490. NASA have sent 2,478 jellyfish into space to see how they would adapt to zero-gravity. The jellyfish mated until there were over 60,000 of them. This sounds like the premise of a horror movie.

Also, isn't sending over 2,000 jellyfish a bit

excessive? Couldn't NASA just send three?

491. NASA have two satellites above Earth tracking the distance between themselves to measure gravitational abnormalities. Since one satellite seems to be chasing the other, they are called Tom & Jerry.

492. NASA wanted to capture Russian satellites with their first space shuttle. This plan never came to fruition.

493. You'd think it doesn't take a rocket scientist to invent something as a simple as a Super Soaker. However, that's exactly what happened. The Super-Soaker was invented by a rocket scientist called Lonnie Johnson in 1982. Johnson created this device as a substitute to pump heat into a refrigerator. Realizing that the pump squirted the water too intensely, he thought it worked better as a water gun for children.

494. 3D printers nowadays can print almost anything – cars, houses, 3D printers, smartphones, and human organs. NASA's own Jet Propulsion Lab personnel are hoping that they will be able to develop a 3D printer that can build a spacecraft, changing

space exploration forever.

495. NASA technicians eat peanuts for good luck before a space probe reaches its destination.

496. NASA developed a technology called FINDER which can detect heartbeats through 30 feet of solid rock or 20 feet of solid concrete. It has been used to find people during natural disasters and was invaluable during the 2015 Nepal earthquakes.

Space Exploration

497. Yuri Gagarin was the first man in space. He performed a 108-minute space flight in 1961. After he returned from space, Gagarin was seen as a hero and so, made public appearances around the world. When he visited Manchester, it was pouring rain. Although his convertible had a roof, Gagarin refused to use it because "if all these people have turned out to welcome me and can stand in the rain, so can I."

498. Sputnik 1 was the first artificial Earth satellite. It was launched by the Soviet Union on October 4th 1957 and was in space for 92 days. It was only 23 inches in diameter and weighed 184lbs. It was a polished metal sphere with four external radio antennae that allowed it to broadcast radio pulses. It took 96 minutes for Sputnik 1 to orbit Earth.

499. By the time Sputnik crashed into Earth on January 4th 1958, it had travelled over 40 million miles and went around the globe 1,440 times.

500. Russian leader, Nikita Khrushchev, wanted to show his nation's power by

building a missile called the R-7 that could strike America. He had Sergei Korolev create the weapon, believing it would make Russia seem superior than the United States. Luckily, Korolev believed Russia could show its technological superiority over the US without building a weapon. It was his idea to create the satellite that became Sputnik 1.

501. Sputnik 1 was the defining moment of the Space Race; a period during the Cold War where Russia and America were competing against each other to see who could triumph in space technology. The Space Race took place between 1955-1972.

502. On January 2nd 1959, the Soviet satellite, Luna 1, was the first spacecraft to successfully reach the Moon's vicinity. It was supposed to land on the Moon but it failed. However, it got close enough to be considered a victory for the Soviets over the Americans. It was actually the fifth time the Soviets attempted to crash-land a satellite on the Moon. The previous failed attempts were kept so secret, the Americans had no idea they existed at the time.

503. Astronauts can't cry in space.

504. The first American satellite was Explorer 1. It was launched into space in 1958.

505. In 1959, Explorer 6 took the first picture of Earth. The camera quality wasn't very good back then so the Earth looked like a blurry spoon.

506. The first picture of the Andromeda galaxy was taken all the way back in 1888 by Isaac Roberts.

507. Salyut 1 was the first space station ever. Russia launched this station into space on April 19th, 1971.

508. The US launched their first satellite, SkyLab, in 1973. Although Skylab was supposed to be in orbit for a decade, it only lasted six years.

509. When Skylab crashed into Australia, the Australian government fined NASA $400 for littering.

510. A scientific collaboration called SETI (Search for Extra-Terrestrial Intelligence)

was created on November 20th, 1984. Its purpose is to seek out alien life throughout the universe. One of the main factors in forging SETI was a transmission in deep space in the 1970s. A 72-second transmission was received from Sagittarius 120 light years away in August 1977. It is known as the WOW Signal. You can listen to it on YouTube. To the average listener, the signal sounds like static.

But astronomers have filtered the signal in English, Chinese, French, and Ancient Sumerian and have concluded that it has a distinct pattern. On the other hand, static has no pattern. No one has ever found out what the WOW Signal was.

511. The ISS' Cupola module has the largest window in space, measure 2.95 meters (9ft 8.) It looks a bit like the window on the Millennium Falcon.

512. Sergei Krikalev has spent more time in space than any other person. He was involved in six flights and has spent 803 days 9 hours and 39 minutes in total in space.

513. It's impossible to whistle in a spacesuit.

514. 0.000007% of Earth's inhabitants have flown into space.

515. Due to the low gravity in space, astronauts have great difficulty telling when they need to go to the toilet.

516. Did you ever dream of being an astronaut? Well, guess what? It's reeeeeeeally hard. The average astronaut suffer from a damaged cardiovascular system, muscle atrophy, insomnia, rage, depression, psychosis, and bone deterioration, mainly in the neck and spine.

Clean water is scarce which leads to bacterial infections. The smallest illness can be lethal in space because astronauts' immune systems are heavily weakened.

Spending too much time in space can make an astronaut's heart become spherical in shape. 12 astronauts were tested before and after space missions and ultrasound tests verified that their hearts had become more spherical by 9.4%.

After several months in space, muscle mass can decrease by up to 20%.

Astronauts can lose up to 1.5% of bone density every month.

Also, they usually lose their fingernails after performing a spacewalk.

On top of that, astronauts only work for 6.5 hours per day because space makes them so cognitively impaired, that they can't risk making critical errors.

517. Astronauts get so used to objects floating in space, they let go of objects in mid-air when they return to Earth, assuming they will float.

518. Astronauts can only eat certain foods which means that diet limitations tend to cause a lack of energy in space travellers.

519. Since there's very little gravity in space, the blood in an astronaut's body doesn't move around properly and builds up in the chest and head. This is why astronauts usually have swollen faces.

520. What countries would you associate with space? Most people would say Russia and America. Germany is also a worthy contender for space exploration.

However, many would be surprised to know that India has successfully launched space probes. The Indian Mars Orbiter,

Mangalyaan, has sent astonishingly clear images of Mars.

What's more astounding is that this expedition only cost $74 million. That's really low by space standards. America's last Mars mission, MAVEN cost $672 million. Even the space movie, Gravity, cost more than India's expedition.

521. Over 17,000 items of space junk are bigger than a mug. That doesn't sound too bad… except for one thing…

522. Some of the debris circulating Earth is moving up to 17,500 miles per hour.

523. In 2015, Chris Hadfield released the first album of songs recorded in space.

524. In 2018, the James Webb Space Telescope was launched. One of its tasks was to detect cholroflurocarbons, which is a fancy word for "alien pollution." This sounds absurd but it makes sense. Pollution is created by intelligent life. If this telescope can detect pollution, it can direct us towards extra-terrestrials.

525. Astronauts see about 15 sunrises and

sunsets every day.

526. In 1984, the waste-dump system on the Discovery space shuttle malfunctioned, causing the urine to freeze, which clogged the nozzle. The astronauts were worried that it could cause damage to the shuttle upon reentry and so, spent over three days trying to remove it. That's right. The astronauts greatest concern on this journey was an icicle made of urine.

527. John Young smuggled a sandwich on-board the Gemini 3 mission. Crumbs broke off the sandwich, which could've potentially got stuck in electrical panels, which would have jeopardized the entire mission. Imagine that headline – Astronauts Killed By Sandwich.

528. There is more debris in our orbit every year. Luckily, NASA have figured out a way to counter this. In 2021, astronauts will be going fishing... in space. Astronauts in spaceships will use nets to ensnare space junk, to avoid it colliding into satellites. Scientists are currently building other material to eradicate space debris including

an ion beam, a robotic arm, and a space harpoon. It sounds like Moby Dick in Space.

529. In August 2015, astronauts ate food that was grown in space for the first time. They had a salad with olive oil and balsamic vinegar.

530. Almost every photograph you've ever seen of space has been colored in because the Hubble telescope doesn't have a color camera. It takes black-and-white pictures, which are colored in afterward. Often, they're not even colored in accurately. They are only colored in to highlight certain qualities or to look more interesting.

531. The longest time that any person has spent in space in one stay was 14 months by Valeri Polyakov.

532. Valentin Lebedev spent 211 days in orbit. During that time, he was exposed to high radiation levels and lost his eyesight to cataracts.

533. Russians call their astronauts "cosmonauts."

534. The ashes of 24 space pioneers (including the creator of Star Trek, Gene Roddenberry,) were sent into orbit on April 21st 1997 on an American air-launched rocket called Pegasus.

535. The following animals have been in space - guinea pigs, newts, mice, frogs, fish, turtles, cats, spiders, insects, monkeys, jellyfish, and dogs.

536. Birds can't be sent into space because they need gravity to swallow food. The only bird that doesn't rely on gravity to eat is the dove.

537. Kirobo was the first talking robot in space. He was sent to the International Space Station on August 9th 2013. He speaks Japanese and keeps records of the conversations he has with astronauts.

538. Jim Voss and Susan Lehms spent eight hours and 56 minutes performing duties outside of their shuttle. It is considered to be the longest time anyone has spent working consistently outside a space shuttle.

539. On August 20th 1977, NASA launched

Voyager 2 into space. Voyager 2 is a spacecraft designed to explore Jupiter, Saturn, Uranus, and Neptune.

540. It took almost four months for Voyager 2 to reach the Asteroid Belt from Earth. It arrived in the Belt on December 10th 1977 and exited it on October 21st 1978.

541. Voyager 2 arrived at Jupiter on April 25th 1979.

542. Voyager 2 arrived at Saturn on June 5th 1981.

543. Voyager 2 arrived at Uranus on November 4th 1985.

544. After a decade of space exploration, Voyager 2 was between Uranus and Neptune.

545. Voyager 2 arrived at Neptune on June 5th 1989. This means it took 12 years for Voyager 2 to reach the furthest planet from Earth in the Solar System.

546. Voyager 2 will run out of power in 2025.

547. New Horizons was launched into space on January 19th 2006. It reached Pluto on July 14th 2015. It was the first time a spacecraft has taken a picture of the surface of Pluto.

548. New Horizons has travelled over three million miles.

549. New Horizons had the fastest lift-off of any spacecraft ever. It was on board the Atlas V rocket when it launched at 36,250 miles per hour on January 19th 2006.

550. In the movie, Gravity, a space station crashes and causes a cascade of debris. This is a genuine danger known as the Kessler Syndrome (named after NASA scientist, Donald Kessler.) Kessler theorized that if a space station crashed in space, it would trigger a catastrophic chain-reaction of debris that could damage many satellites, which could affect global communications and television. This would make it impossible to launch other satellites into space.

551. If an astronaut was to have contact with an alien in any way, the astronaut would be quarantined according to the US Congress'

1969 Extra-terrestrial Exposure Law.

552. In 2015, the Russian space program announced plans to send an all-female crew on an exploratory mission. Most commentators at the press conference asked how the women would manage without makeup.

553. In the late 1950s and early 1960s, the US government considered a space operation called Project Orion. This would involve a spaceship being powered by nuclear explosions. Such an idea could theoretically get a ship from Earth to Mars in weeks. The idea was abandoned in 1964.

554. Scientists at a radio telescope dedicated 20 years to find the source of signal interference they kept picking up, believing it could be a message from extra-terrestrial life. They eventually learned it came from their microwave.

555. It would cost $12.4 billion to build a Starship Enterprise with the same dimensions as the ship in Star Trek. However, since it would weigh 228,000 tons, it would take $456 billion to send it into

space.

556. Russian cosmonauts take guns into space in case they land on Earth off-course and have to defend themselves against bears.

557. The New York Times said, "A rocket will never be able to leave the Earth's atmosphere" in 1936. I hope whoever said that got fired.

558. 20 years ago, an electromagnetic propulsion drive (or EM drive) was built but was dismissed after it was deemed impossible to use effectively for space travel.

As technology has advanced over the years, this EM drive is now being considered for interstellar travel. With this drive, it would only take four hours to travel to the Moon, 70 days to get to Mars, and 18 months to land on Pluto.

This drive would allow humanity to visit Proxima Centauri (the nearest star apart from the Sun) within a century. With current technology, it would take tens of thousands of years to reach it.

559. Observatories have to be built as far away from cities as possible since artificial

light from signs, street lamps, and buildings can hinder astronomers' ability to see space bodies in the sky.

560. Although people think of NASA when they think of space explorations, there are 71 space facilities throughout the world.

561. As Voyager 2 turned around, it took a picture of Earth. This picture is known as The Pale Blue Dot. Astronomer, Carl Sagan, said this about the picture –

"We succeeded in taking that picture, and, if you look at it, you see a dot. That's here. That's home. That's us. On it, everyone you ever heard of, every human being who ever lived, lived out their lives. The aggregate of all our joys and sufferings, thousands of confident religions, ideologies and economic doctrines, every hunter and forager, every hero and coward, every creator and destroyer of civilizations, every king and peasant, every young couple in love, every hopeful child, every mother and father, every inventor and explorer, every teacher of morals, every corrupt politician, every superstar, every supreme leader, every saint and sinner in the history of our species, lived there on a mote of dust, suspended in a sunbeam.

The earth is a very small stage in a vast cosmic arena. Think of the rivers of blood spilled by all those generals and emperors so that in glory and in triumph they could become the momentary masters of a fraction of a dot.

Think of the endless cruelties visited by the inhabitants of one corner of the dot on scarcely distinguishable inhabitants of some other corner of the dot.

How frequent their misunderstandings, how eager they are to kill one another. Our posturing, our imagined self-importance, the delusion that we have some privileged position in the universe, are challenged by this point of light.

Our planet is a lonely speck in the great enveloping cosmic dark. In our obscurity, there is no hint that help will come from elsewhere to save us from ourselves. It is up to us. It's been said that astronomy is a humbling, and I might add, a character-building experience. To my mind, there is perhaps no better demonstration of the folly of human conceits than this distant image of our tiny world.

To me, it underscores our

responsibility to deal more kindly with one another and to preserve and cherish that pale blue dot, the only home we've ever known."

Exoplanets

562. An exoplanet is a planet that orbits a star other than the Sun. The first exoplanet was discovered in 1992.

563. In 1995, the first exoplanet orbiting a star like the Sun was discovered. Michel Mayor and Didier Queloz learned of the planet and called it 51 Peg B. It orbits its star, Peg.

564. A year on 51 Peg B is 4.23 Earth days.

565. 51 Peg B is only five million miles away from its star. That's about seven times closer than Mercury is to the Sun.

566. 51 Peg B is about half the mass of Jupiter.

567. Discovering 51 Peg B changed the fundamental understanding of planets. Before its discovery, astronomers believed it was impossible for a planet to be that big when it is that near a star. It turns out that they were right. The planet formed much further away and migrated to the star. Big planets that migrate towards their star like

the way 51 Peg B has are known as Hot Jupiters. Most of the first few exoplanets discovered were Hot Jupiters. Oddly, these planets were not considered irrefutable proof that exoplanets existed.

One day, that all changed...

568. In 1999, a planet known as HD209468b (which is weirdly similar to my Internet password) was discovered. While observing the planet's star, its light dramatically dropped and then suddenly re-emerged. The only explanation why this would happen is if a planet travelled past it, blocking its light. This is known as a transit. Scientists calculated how big the planet was and how fast it was travelling. HD209468b became the first official exoplanet in the cosmos.

By locating other transits, hundreds of exoplanets were found. Transits are now considered the easiest techniques to locate new planets.

569. In 2004, the first photograph of an exoplanet was unveiled. Its name was 2M1207b... I'm just going to call him Bob.

Bob has five times more mass than Jupiter. It was discovered because its galactic system is very young so it is bright enough to appear in a telescope and hot enough so it appears in infrared photographs.

570. Between January-July 2015, 521 exoplanets were found. That means that two or three exoplanets are found every day.

571. On February 2015, the Extrasolar Planets Encyclopaedia stated they had found 1,890 exoplanets.

572. The oldest exoplanet is 12.7 billion years old. It's called PSR B1620-26 b (but it's better known by its nickname, Methuselah.) It was discovered in 1993 and formed about a billion years after the universe began.

573. PSR J1719-1438 b is a planet that is made of diamond. It is the densest planet that astronomers have ever found.

574. The nearest exoplanet is called Epsilon Eridani b. It is only ten light years away.

575. The exoplanet with the longest year is called 2MASS J2126-8140. A year on this planet is approximately 900,000 years long.

576. Sweeps 04 is the furthest planet ever discovered. It is 27,710 light years away.

577. Tres 2b is an exoplanet that is nicknamed the Dark Planet because it reflects absolutely no light and is entirely black.

578. HD 100546 b is the largest exoplanet astronomers have found. It is 6.9 times larger than Jupiter.

579. Although HD 100546 b is the biggest known planet, another planet called DENIS-P J092303.1-491201 b has the most mass. It has 28.5 times more mass than Jupiter.

580. Gliese 436 b is a unique exoplanet because it may be covered in burning ice.

581. A planet called Kepler-452b has been discovered 14,000 light years away. It resembles Earth more than any planet in the known universe.

582. Kepler-452b is more commonly known as Earth's twin or Earth 2. NASA employees call it Coruscant, which is the name of a planet in the Star Wars franchise.

583. Earth 2 is 60% larger than our planet and has about five times more mass.

584. Earth 2 is nearly the same distance from its star as our planet is from its Sun.

585. Earth 2's year is 384 Earth days long.

586. Although astronomers are unsure what Earth 2's atmosphere is like, it seems very likely that it harbours liquid water.

587. Earth 2's star is 1.5 billion years older than the Sun.

588. The gravity on Earth 2 is twice as strong as ours so a person would weigh double what they weigh on Earth.

589. Although Earth 2 sounds amazing, don't get too excited. With current technology, it would take 28 million years to get there.

590. The smallest exoplanet found is Kepler 37b. It's smaller than any planet in our Solar System. It's slightly bigger than the Moon.

591. Hundreds of Earth-sized planets have been found so far.

592. Any exoplanet that is larger than Earth but smaller than Uranus or Neptune is known as a Super Earth.

593. It has been calculated that there are approximately ten billion Earth-like planets in our galaxy.

594. So far, 500 star systems similar to the Solar System have been found, each with many exoplanets.

595. The most abundant star system ever discovered has seven planets orbiting its star.

596. Planets can orbit more than one star. Exoplanet PH1b has a complex orbit around two stars that orbit each other. Weirder still, these three orbit around another pair of stars that are orbiting each other.

597. Exoplanets have been found around almost every kind of star – Lower Mass Stars, High Mass Stars, Red Dwarfs, Red Giants, Blue Stars, etc.

598. In 2014, almost 800 planets were discovered. Since records started in 1995, between 20-100 planets were discovered per year. That means that in 2014, more planets were discovered than in the previous 20 years.

599. Kepler 7b is a planet that is 1.5 times larger than Jupiter, it is as dense as Styrofoam.

600. The planet 55 Cancri E has the fastest change of temperature in the cosmos. This planet's temperature changes from 1,000-2,700 degrees Celsius in 18 hours.

601. The surface of the exoplanet, KELT-9b, is 4,315 degrees Celsius, making it the hottest known planet in the cosmos. In fact, it's hotter than some stars.

602. Because of the volume of exoplanets uncovered in recent years, astronomers have calculated that there are hundreds of billions of planets in our galaxy alone.

Stars

603. The Greek astronomer, Hipparchus of Nicaea, was the first person to create a star catalogue. He ranked them by what he called "Magnitudes." The brightest stars were First Magnitude, the next brightest stars were Second Magnitude, etc. Astronomers of today use a similar system to catalogue stars.

604. Stars are made of the following elements in descending order – Hydrogen, helium, oxygen, carbon, nitrogen, silicon, magnesium, neon, iron and sulphur.

605. A star is 91.2% hydrogen in matter.

606. A star is 71% hydrogen in mass.

607. Stars can be red, yellow, white, or blue.

608. Astronomy means "law of the stars."

609. Stars are usually born in groups called clusters. Stars rarely form by themselves.

610. Clusters can form hundreds of thousands of stars at a time.

611. Galileo was the first person to identify clusters.

612. The only two clusters that are visible to the naked eye are The Pleiades and Hyades.

613. The best time to see Pleiades is 4am in September, 12am in November, or 8pm in January.

614. Pleiades is in the constellation of Taurus.

615. There are two types of clusters – Globular Clusters and Open Clusters.

616. Open Clusters are loosely bound collections of dozens or thousands of stars. Open Clusters are also known as Galactic Clusters.

617. Around 1,100 Open Clusters are known today.

618. Open Clusters are only a few hundred million years old, which is very young by astronomical standards.

619. A Globular Cluster is a swirling orb of stars. These types of Clusters have far more stars than Open Clusters.

620. Globular Clusters can be hundreds of light years across.

621. The nearest Open Cluster is 440 light years from Earth.

622. Stars in clusters don't orbit each other systemically like our Solar System. Their orbits zip around in all directions, like bees in a hive.

623. Because of the erratic nature of stars in a Globular Cluster, stars can collide. When this happens, they merge and become a Blue Strangler.

624. Globular Clusters are around ten billion years old.

625. The nearest Globular Cluster is Messier 4. It is 7,000 light years from Earth.

626. Ptolemy of Alexandria spotted a Globular Cluster nearly 2,000 years ago.

627. Globular Clusters were the first objects to fully form after the universe was created.

628. There have been 150 Globular Clusters identified in the Milky Way.

629. Globular Clusters probably don't have any planets orbiting stars inside them.

630. If a star is less than 300 million years old, it is considered to be a "new star." These young stars are called Cepheids. Their brightness fluctuates rapidly, which makes it easy to find them.

631. Many people believe Polaris the North Star is the brightest star in the sky but this is incorrect. The brightest star in the sky is Sirius the Dog Star.

632. The smallest known star in our galaxy is VB 10. It is only 20% larger than Jupiter.

633. In 2016, NASA's Kepler and Swift missions found 18 stars that spin so fast, they have squished their shape into ovals. They have been nicknamed Extreme Pumpkin Stars. These types of stars only take two or three days to make a full

rotation. By comparison, the Sun takes almost a month to rotate and it's much smaller than these stars. Extreme Pumpkin Stars spin so fast, they spew solar flares, x-rays, and radiation in every direction for millions of miles.

634. The largest known star is UY Scuti. It is 1,700 times wider than the Sun and is 1.47 billion miles across.

635. UY Scuti is so big that if it was in the same spot as the Sun, it would consume every planet in the Solar System apart from Saturn, Uranus, and Neptune.

636. The Hyper Giant, Pistol, is 1.6 million times brighter than our Sun.

637. Astronomers have witnessed a star being born. The star, VLA2, was first spotted in 1996 by the VLA radio observatory in New Mexico. At the time, it was simply a dense cloud of gas. As the years went by, astronomers saw the cloud turn into a star. In 2014, it looked drastically different. This means that astronomers have seen the first 18 years of a star's life. Considering our star

is nearly five billion years old, this is astounding.

638. When stars die, they collapse into a Black Hole, a White Dwarf, or a Neutron Star.

639. Only very large stars can collapse into a Black Hole. Smaller ones collapse into Neutron Stars. Average-sized stars like our Sun collapse into White Dwarfs.

640. The nearest star system to Earth after the Sun is Alpha Centauri. It is 4.2 light years away.

641. Alpha Centauri is composed of three stars. The closest of these stars to Earth is Proxima Centauri.

642. Alpha Centauri was discovered in 1893.

643. If you have perfect 20/20 vision, almost 2,500 stars are visible in the night sky with the naked eye.

644. When Europe was going through the Dark Ages, a Persian astronomer called Abd ar-Rahman as-Sufi (or Azophi for short)

discovered many stars. Unsurprisingly, the first stars to be discovered were the most visible. This is why the brightest stars have Arabic names.

645. If you asked a random person to name a constellation, he or she would probably say the Big Dipper. However, that is only a part of a constellation. The Big Dipper makes up a small part of The Great Bear Constellation, Ursa Major.The Dipper makes the Bear's tail...even though bears don't have tails. (How do people figure out how stars connect to each other but they can't double-check to see if a bear has a tail or not?)

646. A star performs nuclear fusion. This is a nuclear reaction that fuses two or more smaller atoms into a larger one, releasing a huge amount of energy.

647. There is an easy way to tell if a celestial body in the sky is a star or a planet. If it seems to twinkle, it's a star. If it doesn't, it's a planet. Speaking of twinkling...

648. Stars don't really twinkle. It is an illusion caused by the movement of the Earth's atmosphere.

649. The oldest living stars are Red Dwarfs.

650. When a star begins to die, it will swell up until it becomes a hundred times its original size. When it does this, it is known as a Red Giant.

651. A Red Giant will eventually collapse into what is known as a White Dwarf. A White Dwarf is a star at the very end of its life. It's very dense but also quite small, (about a billion times smaller than its original size.) This would make it the same size as Earth with 1% of its original diameter.

652. When a star becomes a White Dwarf, it becomes so dense that one centimetre of a White Dwarf (about the size of a dice) would weigh a ton.

653. The gravity of a White Dwarf is a hundred thousand times stronger than Earth's. This means that a normal human would weigh 7,500 tons. That's the same size as a Stealth Destroyer battleship.

654. White Dwarfs have a temperature of 100,000 degrees Celsius.

655. Over 10,000 White Dwarfs have been found in our galaxy alone.

656. Although White Dwarfs are super-hot, they are not very bright because they are so small. This makes it very difficult to spot them.

657. When two stars collide, the resulting explosion is known as a Red Nova.

658. Cecilia Payne-Gaposchkin is one of the most influential people in helping astronomers understand how stars work. For centuries, scientists believed the Sun was composed of similar material to the Earth. In 1925, Payne-Gaposchkin correctly concluded that the Sun was a ball of gas made almost entirely of hydrogen and helium. She was only 25 at the time.

659. Payne-Gaposchkin made over three million star observations in her lifetime.

660. In Star Wars, Luke Skywalker is seen on his home planet of Tatooine, gazing at a double sunset. Although it seems like a sci-fi fantasy to have a planet orbit two stars

simultaneously, such a planet has been found. The planet is called Kepler-47c (but who is going to call it anything except Tatooine?) The planet circles what is known as binary stars. The planet is classified as an icy Gas Giant.

661. The first binary star was discovered thousands of years ago. There is a light in the Big Dipper constellation that looks like a star. However, if you look closely, you will see two stars. This was first observed by Ancient Greeks.

662. The binary stars in the Big Dipper are called Mizar and Alcor. Identifying Mizar and Alcor was an eye test in ancient times.

663. In recent times, telescopes have gotten so advanced, that astronomers learned that Mizar and Alcor were not binary stars; they are sextuple stars! This means that six stars were all caught in each other's gravity.

664. If a star system has more than two stars, it is known as a Multiple Star System.

665. Another Multiple Star System is the North Star, Polaris. Although many people

consider Polaris to be a single star, the reason why it is so bright is because it is a pentuple star system, which means it is made of five stars.

666. Stars make energy in their core by fusing hydrogen into helium.

667. Some binary stars can orbit each other in days.

668. Other binary stars can take centuries to orbit each other.

669. The binary system, 4U 1820-30 orbits each other in 685 seconds, making it the fastest binary star orbit ever recorded. That's 11.4 minutes. It takes me longer to make toast.

670. When binary stars become too close to each other, they drastically speed up and their gravity hurls each other away.

671. Stars in Multiple Star Systems have to flow in each other's gravity perfectly or they could collide or hurl one another into the cosmos. For all we know, the Sun could have had multiple star siblings which were hurled

into space billions of years ago and the Sun is simply the last star standing.

672. Sometimes, binary stars connect to each other without crashing into each other. They don't exactly merge as they maintain their shape. (Think of it like a cosmic conjoined-twin.) These are called Contact Binaries. Their overall shape varies depending on what kind of stars they are.

673. Earlier, I mentioned that some astronomers believed that Jupiter is a "failed star" and I dismissed this idea because it's too small. However, it is possible for planets to turn into stars if they absorb huge amounts of mass (at least 65 times the size of Jupiter.)

674. Turning a planet into a star is not an easy process. When a planet absorbs enough mass, it shrinks until it's about twice the size of Jupiter and cools down and becomes charred and black. This is called a Brown Dwarf.

675. Brown Dwarfs are never brown. I don't know why the name stuck.

676. It rains molten iron on Brown Dwarfs.

677. Shiv Kumar was the first astronomer to theorize the possibility of Brown Dwarfs in 1963.

678. The first Brown Dwarf was called Teide 1 and was discovered in 1995.

679. Brown Dwarfs get mistaken as planets and stars all the time, even by professional astronomers.

680. Brown Dwarfs' light can only be seen in infrared.

681. When Brown Dwarfs absorb matter, they become heavier but they don't become bigger.

682. Over 2,000 Brown Dwarfs have been discovered over the last 20 years.

683. Stars can shoot huge charges of concentrated energy known as solar flares. Solar flares are so powerful, they can unleash 10% of the entire output of a star.

684. There are two classes of star – Low Mass Stars and High Mass Stars.

685. A High Mass Star has at least eight times the mass of the Sun.

686. The lower the mass of a star, the longer it will live.

687. A Lower Mass Star like the Sun can live for about ten billion years. A High Mass Star will live for two billion years. Despite the fact that a High Mass Star has more energy, it uses it up more quickly since its core is under more pressure.

688. Red Dwarfs are the most common type of star in the known universe.

689. A Red Dwarf can live for… over a trillion years. That's 1,000,000,000,000 years. Just to remind you, the universe isn't even 14 billion years old yet. That means that a Red Dwarf can live about 71 times longer than the universe's age today.

690. The hottest stars are Blue Stars. Their surfaces are at least 6,000 degrees Celsius.

691. Red Giants are the coolest type of stars. They are about 3,500 degrees Celsius.

692. When High Mass Stars start to die, they become Red SuperGiants. SuperGiants become a thousand times larger than their original sizes,

693. The most famous SuperGiant is Betelgeuse (which is pronounced “Beetle Juice.”)

694. When a High Mass Star dies, it doesn’t fade away like a White Dwarf. Instead, it explodes. This explosion is called a supernova.

695. A supernova creates a shockwave which shoots through space carrying the ejected material with it, and disrupting whatever it hits. The mess that this leaves behind is called a Remnant.

696. A supernova is one of the most powerful forces of energy in the universe. When a star goes supernova, the explosion is brighter than an entire galaxy.

697. The most famous supernova is the Crab Nebula. This nebula was created when a supernova occurred in 1054. Islamic and Chinese astronomers observed and documented the Crab Nebula supernova in 1054. It was so bright, it could be seen in broad daylight.

698. It can take months for the brightness of a supernova to fade.

699. Supernovas can be seen halfway across the universe.

700. The last supernova in our galaxy was in 1604.

701. When a star goes supernova, it will expel more energy in a micro-second than our Sun will in its entire life.

702. A supernova expels material with so much force, that it moves 20,000 miles per second.

703. The last observed supernova was in 1885. This supernova occurred in the Andromeda galaxy.

704. Astronomers have recently found evidence that suggests a portion of a White Dwarf star can survive going supernova. Although these stars are called Type 1A Supernova, they're nicknamed Zombie Stars.

705. Humanity doesn't have to worry about supernovas affecting our planet. A supernova would have to be a hundred light years away to damage Earth

The nearest High Mass Star to Earth is Spica. It's about 260 light years away.

706. If a star is very, very heavy, it will turn into a Neutron Star when it goes supernova.

707. Neutron Stars are only 15 miles across but are so dense, they can be twice as heavy as the Sun.

708. Neutron Stars are not made of liquid, solid, or gas. They are made of a completely different form of matter known as neutronium.

709. Neutronium is so dense, a 1cm dice made of neutronium would weigh 400 million tons. That's about 500 times heavier than the Golden Gate Bridge.

710. A Neutron Star's gravity is 100 billion times stronger than Earth's.

711. If you could stand on a Neutron Star (which you couldn't because you would be busy being dead) you would weigh 8.5 billion tons. By comparison, the Great Pyramid of Giza weighs 6.5 million tons.

712. Neutron Stars were discovered in 1965.

713. Neutron Stars that shoot out beams of radio wave energy are called pulsars. It's like the Neutron Star is a lighthouse and the radio waves are the light beam.

714. Pulsars have so much electromagnetic radiation, that they regularly interfere with telescope technology.

715. The first pulsar was identified on November 28th 1967 by Jocelyn Bell Burnell and Antony Hewish.

716. Quantum astrophysicists speculate that there might be another type of star called a Quark Star. A Quark Star may form when it's not big enough to become a Black Hole but

it's too large to become a Neutron Star. This type of star would not be made of solid, liquid, or gas but of a different type of matter known as Strange Matter (which scientists still don't really understand today.) At this moment, Quark Stars are merely hypothetical.

717. Physicists have hypothesized that there may be a stage between a Quark Star and a Black Hole. This means that there may be a hypothetical star between another hypothetical star and a Black Hole. (Science is weird.) Anyway, this hypothetical star is an Electroweak Star. It would be made of the same matter that existed in the first one-billionth of a second of the Big Bang.

718. When a big star consumes a neighbouring smaller star, it is known as a Cannibal Star.

719. Astronomers have hypothesized the possibility of a star that has a Black Hole as a core. This hypothetical structure is called a Quasi-Star.

720. Scientists are looking for a different type of star called a Thorne-Zytkow Object

(TZO.) A TZO occurs when a Red SuperGiant consumes a Neutron Star. The collision of two incredibly powerful but very different stars would have a unique and fascinating result. However, this has never been observed.

721. The most powerful Neutron Stars are Magnetars. Their name refers to the fact that they have the most powerful magnetic fields in the universe. A Magnetar's magnetic field is one quindrillion (1,000,000,000,000,000) times stronger than the Sun's.

722. Only 10% of Neutron Stars become Magnetars.

723. The first Magnetar was discovered on March 5th 1979.

724. Magnetars are so unstable, they suffer starquakes (It's like an earthquake but on a star.) A starquake can ignite a Magnetar flare. (At this point, it sounds like I'm just naming Transformers.) A Magnetar flare emits as much energy in a micro-second as the Sun emits in 250,000 years.

725. Although the Richter scale was designed

so Magnitude 10 would be the maximum measurement for an earthquake, the highest ever recorded was.... Magnitude 23. Just to give you an idea how powerful that is, each number in the Richter scale is ten times stronger than the previous number. Magnitude 8 is ten times more powerful than Magnitude 7. This means that Magnitude 9 is a hundred times stronger that Magnitude 7. A Magnitude 9 or higher would obliterate an entire area, permanently change the ground's topography, and be felt on the other side of the planet. Magnitude 9 earthquakes happen about once every 30 years and have killed hundreds of thousands of people. So how is a Magnitude 23 possible? That's 100 trillion times stronger than any recorded earthquake. However, this quake happened on a Neutron Star called SGR 1806-20. It was so powerful, it destroyed everything in a 10-light year radius. In 1/10th of a second, this quake released more energy than the Sun does in 100,000 years.

726. As the universe gets older, it will become cooler. This will turn stars cooler and icy and they will be known as Frozen Stars.

727. The largest diamond in the universe is a white dwarf star that measures 2,500 miles across and is situated in the Centaurus constellation. It's been nicknamed Lucy after the Beatles song, Lucy in the Sky with Diamonds.

728. NASA have recently found a star that is destroying its neighboring planet with intense gravity. Although the star has no official name, it has been nicknamed the Death Star.

729. There are more stars in the universe than grains of sand on Earth. However, there are more atoms in one grain of sand than stars in the universe.

730. There are some stars that are cool enough to touch.

731. Although the star, Icarus, died billions of years ago, its light was photographed by a Nature Astronomy team in April 2018. Since it is nine billion light years away, Icarus is the furthest star ever witnessed.

732. The first stars in the universe were very short-lived (only about two million years.) These were called Population III stars.

The Sun

733. The Sun's real name is Sol. It is named after the Roman god of the stars. A "Sun" is a star that has planets orbiting it.

734. The Sun is 4.57 billion years old.

735. The Greeks called the Sun "Helios."

736. The Sun's light will reach the nearest star in 4.3 years.

737. The Sun travels at a speed of 136 miles per second.

738. It would take 52,000 years for the Sun's light to travel to the edge of our galaxy.

739. The Sun has a surface area of 3.7 trillion miles squared.

740. The Sun has a diameter of 945 million miles.

741. The Sun makes up 99.8% of the mass of the Solar System. 0.1% is made up by Jupiter. The other 0.1% if made up of everything else.

742. If you were standing on the Sun (and not melting,) you would be 435,000 miles from its centre.

743. Despite what many people believe, the Sun isn't on fire. Fire can only exist with oxygen which space is lacking. The Sun emits light and heat when its energy is converted into gamma rays.

744. It takes seven hours for the Sun's light to reach the edge of the Solar System.

745. A fear of the Sun is called heliophobia.

746. It would take a regular plane 20 years to reach the Sun from Earth.

747. The Ancient Greek teacher, Ptolemy, taught the Ptolemaic system (also known as geocentrism.) This theory suggested the Earth was at the centre of the universe. This idea was accepted until the 16th century.

748. The idea that the planets revolve around the Sun and our star is the centre of the Solar System is known as heliocentricism.

This idea was popularised by Copernicus, Galileo, and Kepler.

749. Galileo Galilei put forward his theory that the Earth revolved around the Sun in 1632 in his book, Dialogue Concerning the Two Chief World Systems. He was sentenced to house arrest by the Inquisition for his work because it contradicted the notion that the Earth was not the centre of the universe.

750. 1.3 million Earths could fit in the Sun.

751. The Sun's gravity is 28 times stronger than Earth's.

752. The surface of the Sun is 5,800 degrees Celsius.

753. Only three of the 2,500 visible stars in the night sky are smaller than the Sun.

754. The Sun weighs two octillion tons. That's 2,000,000,000,000,000,000,000,000,000 tons.

755. The Sun's diameter is ten times that of Jupiter.

756. The pressure at the Sun's core is 260 billion times stronger than the pressure of Earth's atmosphere.

757. Aristarchus of Ancient Greece was the first person to theorize that the Sun rotated like the Earth. This was in 250 BC when other philosophers believed the Sun was stationary. Sadly, Aristarchus was dismissed.

758. The Sun is very small compared to other stars. It is categorised as a dwarf star, which is the smallest kind of star.

759. The Sun is white. It seems red or yellow because it's constantly exploding.

760. In 1593, a friar and astronomer called Giordano Bruno infamously declared to the Catholic Church that the Earth was not the center of the universe and the Sun was simply one of countless stars. What is most bizarre about this claim is that Bruno didn't have a shred of proof that the Earth revolved around the Sun. Bruno believed in this concept because he saw it in a dream. He simply liked the idea that the Earth wasn't the center of everything because it meant

the universe was far bigger than people believed. He was tried for heresy and was burned at the stake in 1600.

761. Polish mathematician, Nicolaus Copernicus, was the first person to prove that the Earth revolves around the Sun in 1532. Although his theory wasn't perfect, the idea that the Sun was the centre of the Solar System was finally taken seriously.

762. The most significant person in the history of astronomy is the English physicist and mathematician, Isaac Newton. Newton invented calculus, which was instrumental in making calculations of celestial bodies and understanding the Sun's orbit.

763. The first solar flare ever detected was in 1859. It was also the most powerful flare ever witnessed.

764. In 2012, the Sun blasted out a solar flare almost as powerful as the one in 1859. Luckily, it missed Earth.

765. The coolest parts of the Sun's surface are 3,800 degrees Celsius. These areas are known as sunspots. It is not fully understood

what makes sunspots cooler than the rest of the surface.

766. The core of the Sun is 15.6 million degrees Celsius.

767. The amount of energy the Sun expels per second is the equivalent of 400 billion one-megaton nuclear bombs.

768. The Sun expels solar wind at 600,000 miles per hour.

769. The Sun's light starts in the core and takes hundreds of thousands of years before it works its way to the surface, which is the light that we see. This means that the light that you see from the Sun today, started in the core before human civilization existed.

770. Sunspots on the Sun can be bigger than Earth.

771. Some sunspots are so big, they can be seen with binoculars.

772. The Sun hurls massive lumps of flame called Prominences.

773. A Prominence is hundreds of thousands of miles long.

774. Although it's common knowledge that the Moon affects the tides of Earth's oceans, the Sun also affects them.

775. If the Sun lost half of its mass, it would lose its gravitational pull on all the planets in the Solar System and they would stop orbiting the star.

776. The Sun is approaching the middle-age of its life cycle.

777. As the Sun begins to die, the Sun will start to swell and become what is known as a Red Giant. This happens when a star is about 11.5 billion years old.

778. When the Sun becomes a Red Giant, it will absorb Mercury and Venus.

779. When the Sun reaches its maximum size, it will become 2,000 times brighter.

780. Although a Red Giant stage is the biggest our Sun will become, it will lose $1/3^{rd}$ of its mass.

781. As a Red Giant expels most of its energy, it will wither down to a white ball that is about the same size as Earth. This is known as a White Dwarf.

It takes 500 million years for a Red Giant to turn into a White Dwarf.

782. White Dwarfs can't generate energy so it will fade to nothing over a process that takes tens of billions of years. That is how our Sun will die.

783. It takes the Sun 24 days to rotate on its own axis.

784. The oldest known device used to tell the time was a sundial. The oldest sundial was found in the Valley of the Kings in Egypt. It was about 3,550 years old.

785. Chankillo is 2,300 years old, making it the oldest solar observatory in North America, Central America, and South America. It was designed to line up with the sunrise of the summer and the sunrise of the winter.

786. In 1929, a German rocket scientist called Hermann Oberth wanted to create a Heliobeam; a lens that would draw energy from the Sun and redirect it into a single point on Earth causing devastating damage. Thankfully, this was never made (but it did make an appearance in the James Bond movie, Die Another Day.)

Asteroids

787. Asteroid means "star-like."

788. There are countless asteroids located in-between Mars and Jupiter. It is known as The Asteroid Belt.

789. There is an asteroid named after Charlie Chaplin.

790. Nearly any object in the Solar System that didn't become a planet or moon was left to float in the Asteroid Belt. Basically, asteroids are the Solar System's leftovers.

791. There are asteroids across the Solar System but most of them are in the Asteroid Belt.

792. For decades, astronomers didn't know why there was a big gap between Mars and Jupiter. They assumed that there must be a planet hiding there somewhere.

793. Some asteroids have moons.

794. There is more metal in the Asteroid Belt than on Earth.

795. The metal in the Asteroid Belt would be worth at least $100 billion.

796. Asteroids are made of different things –

i) Dark C asteroids are made of carbon.

ii) Bright S asteroids are made of magnesium.

iii) Bright M asteroids are made of iron.

797. Asteroids that orbit planets are called Trojans.

798. An asteroid hit the Earth 1.85 billion years ago. The impact was so powerful, it created a volcanic eruption that lasted a million years.

799. The biggest asteroid impact of recent times was the Tunguska Event in Krasnoyarsk, Russia in 1908. This asteroid struck the Earth at 7.14 am on June 30th. The impact was as powerful as the atomic bomb that was dropped on Hiroshima during World War II. Luckily, the asteroid hit a forest so no one was killed. However, the impact destroyed 80,000 trees.

800. The scariest asteroid that we have record of is the Chiling-Yang. In 1490, an asteroid burst open in China while it entered the Earth's atmosphere. Instead of burning up like a normal asteroid, it split into hundreds of meteorites. According to the Ming Dynasty records, up to 10,000 people died.

801. Earth-striking asteroids are becoming less frequent over time.

802. The average temperature of a typical asteroid is -73 degrees Celsius.

803. Some asteroids are blown-off chunks of comets.

804. Sometimes, asteroids are referred to as minor planets or planetoids.

805. Most asteroids aren't solid. They are just clumps of rubble barely kept together by gravity.

806. NASA's Galileo was the first spacecraft to take a close-up image of an asteroid in 1991.

807. NASA's Galileo discovered the first asteroid moon in 1994.

808. In 2001, NASA's spacecraft, NEAR, became the first spacecraft to land on an asteroid when it settled on Eros. What's more amazing is that NEAR was never designed for landing.

809. In 2006, Japan's Hayabusa became the first spacecraft to land on and take off from an asteroid. It returned to Earth in June 2010. Its collected samples are still being studied to this day.

810. Asteroid fields are never clustered. They are so sparse, that a spacecraft could travel through the Asteroid Belt without coming into contact with a single asteroid.

811. Each asteroid in the Asteroid Belt is about three million miles away from the nearest asteroid.

812. An asteroid killed off the dinosaurs when it collided into Earth 65 million years ago. The asteroid is called Chicxulub.

813. Chicxulub's impact was so powerful, it created debris that was so big, it blocked out the Sun worldwide for many years.

814. Chicxulub slammed into Mexico's Yucatan Peninsula.

815. The Yucatan crater is 110 miles wide.

816. For decades, no one knew what killed off the dinosaurs. But in 1980, Luis Alvarez and Walter Alvarez discovered an impact crater near the Yucatan coast of Mexico. Since then, the dinosaur-killing asteroid theory has been considered the only logical explanation to the death of the dinosaurs.

817. Every few million years, an object (meteor, asteroid, comet) big enough to threaten Earth's civilization impacts our planet.

818. NASA are considering building a net to capture potentially dangerous asteroids. The net is called Weightless Rendezvous and Net Grapple to Limit Excess Rotation. It's more commonly known as WRANGLER.

819. Most asteroids are 75% carbon based.

820. If you added up all the mass of the asteroids, it would only be 4% of the mass of the Moon.

821. Asteroids are almost never spherical. They are normally ellipsoids.

822. Computer models show that an asteroid called Baptistina smashed its neighbouring asteroids 160 million years ago. This collision caused hundreds of asteroids to be ejected from the Asteroid Belt and be hurled into space... in the direction of Earth. It is incredibly likely that Baptistina began the chain reaction that cast an asteroid towards Earth, which killed off the dinosaurs.

823. The largest asteroid in the Asteroid Belt is Ceres. It has a diameter of 590 miles. It's about the same size as Texas.

824. Ceres makes up 1/3rd of the weight of the Asteroid Belt.

825. The Dawn satellite did a flyby of Ceres and detected that it has a cryovolano, which means that it has an underground ocean.

826. Ceres was the first asteroid to be discovered.

827. When Ceres was first discovered, astronomers thought it was a comet.

828. Ceres is one of the only asteroids to be spherical.

829. Because of its shape, Ceres is considered to be a dwarf planet as well as an asteroid.

830. Two asteroids are named after former president, Herbert Hoover – Herberta in 1935 and Hooveria in 1920.

831. Over a million asteroids are over one kilometer across.

832. Most asteroids are about a hundred meters across.

833. Most asteroids are named after Greek and Roman gods and goddesses.

834. The Asteroid Belt probably has over a billion asteroids.

835. Over 440,000 asteroids have been identified in the Asteroid Belt.

836. The more stone that is in an asteroid, the more likely that it will shatter if it breached Earth's atmosphere, no matter how big it is.

837. The Kleopatra asteroid is shaped like a dog bone.

838. The Toutatis asteroid is shaped like a dumbbell.

839. Once per year, a car-sized asteroid hits Earth's atmosphere. It usually burns up before it hits the ground.

840. On November 10th 2013, Hubble telescope spotted an asteroid with six tails.

841. There are some asteroids that could potentially veer away from the Asteroid Belt and head in Earth's direction. These are known as Near Earth's Objects (NEO's.) NASA monitor NEOs to make sure the potential threat of colliding into Earth is minimal.

842. Four NEO's are catalogued by NASA.

843. The most dangerous NEO is called Apophis. In the year, 2029, Apophis will come very near Earth. Experts say it has a 2.5% chance of colliding with our planet.

Meteorites

844. A meteoroid is a small rock in space.

845. A meteor is a meteoroid that burns up in Earth's atmosphere. It is more commonly known as a shooting star.

846. Meteoroids have been found on the Moon.

847. A meteorite is a space rock that survives passing through Earth's atmosphere and falls to the ground.

848. When a meteor passes through the Earth's atmosphere, it can move at 45 miles per second.

849. Up to a hundred tons of meteors hit the Earth every day.

850. A meteor can become so fast and so bright, it can create a streak in the sky called "a train."

851. A meteor's train can be visible for up to 30 minutes.

852. Meteoroids can be dangerous because they are too small to be seen with current technology.

853. There are three types of meteorites – Stony, Iron, and Stony Iron. The most common type of meteorite is Stony.

854. 90% of all meteorites found on Earth were found in Antarctica.

855. The oldest meteorite ever found is 4.568 billion years old. That's as old as the Solar System.

856. A meteor that is as bright as a full Moon is classified as a Bollide.

857. "Bollide" means "missile" in Greek.

858. Sometimes, you can see many meteors hurtling through the sky. These are called meteor showers.

859. The most famous meteor shower is the Perseids. In the Perseids, you can see about a hundred meteors every hour.

860. Meteorites have been found on Mars.

861. The Perseids were the first meteor shower ever seen. They were recorded in China in 36 AD.

862. Most meteor showers occur in December, especially in the Northern hemisphere. However, the Perseids are visible from mid-July to late August.

863. The best time to see a meteor shower is 3-4am on a moonless night.

864. Most meteors in a meteor shower are caused by the debris of comets.

865. The meteors that you see in a meteor shower are about 30-40 miles above the Earth's surface.

866. Very powerful meteor showers can produce over a thousand meteors an hour. These are known as meteor storms.

867. The Meteor Data Centre lists about 600 meteor showers.

868. About 500 meteorites reach the Earth's surface per year. However, only five of them

will be collected by scientists to study. Many are lost in the ocean.

869. A lot of meteors fall through the atmosphere during the day. Sadly, we can't see them because it's too bright.

870. In 2004, a 30ft meteorite struck the Earth, creating two million pounds of dust. Luckily, this happened in Antarctica so nobody was hurt. However, the force was so powerful, that it affected the climate on the other side of the planet.

871. If a meteor is found, it is called a Find.

872. If a meteor is collected by a scientist, it is called a Fall.

873. To date, there has been a thousand collected Falls.

874. To date, there have been 40,000 Finds.

875. Of the 40,000 Finds, 34 of them originated from Mars.

876. Meteorites are named after the places they are found in. If two meteorites land in

the same place, they end up with the same name. FOREVER.

877. Every 2,000 years, a meteorite the size of a football field hits the Earth and causes significant damage.

878. Although meteorites are radioactive, the radiation is too faint to pose a threat to humans.

879. Most meteors are 25 metres wide. These usually burn up in the atmosphere and won't be seen by the naked eye.

880. A meteor larger than 25 metres wide is likely to cause local damage to the area.

881. If a meteor is larger than 1.4 miles, it can have cataclysmic effects on a global scale.

882. Four billion meteoroids fall to Earth daily. Most are too tiny to do any harm.

883. The International Space Station is encrusted with a foot-thick layer of Kevlar. Kevlar is the material used to make bullet-proof jackets. The Kevlar protects the space

station as it has been hit by 100,000 meteoroids over its 20-year life span.

884. If you find a meteorite, the Nomenclature Committee of the Meteoritical Society will demand that you donate 20% of the rock for research. You can keep the rest.

885. If you find a meteorite in South Africa, you must surrender it to the nearest authority.

886. The most devastating meteorite in recent times was in February 15th, 2013 when a 19-metre-wide meteor entered Chelyabinsk; a Russian city with a population of 1.13 million. Upon the meteor's entry, it shattered into many pieces, creating a force of energy as powerful as half a million tons of TNT. That's nearly as powerful as an atomic bomb. No one died but over a thousand people were injured by glass shattering from the force of the meteorite.

887. Meteorites are among the coldest things in the universe. They are usually -270 degrees Celsius. When a meteorite passes through Earth's atmosphere, it becomes

very hot. However, the core of the meteorite remains freezing cold. As the meteorite passes through Earth's atmosphere, the core will revert the meteorite's surface to a freezing temperature within minutes. A meteorite is so cold, that a person can easily suffer frostbite if he or she touches it.

888. The largest meteorite found on Earth was in Namibia in 1920. It's called Hoba and measure 2.7 metres wide, 2.7 metres deep, 0.9 metres high, and weighed 60 tons.

889. The Fukang meteorite is 4.5 billion years old and weighs one ton. It was found in China and landed on Earth when our planet was only a hundred million years old.

890. When a meteoroid is larger than ten metres in diameter, it is considered an asteroid.

891. On February 6th, 2016, a bus driver in India was killed by a meteorite. This is the first and only recorded time that someone has been killed by a meteorite.

Comets

892. A comet's tail is formed by solar wind. No matter what direction the comet is moving in, the tail will always face away from the nearest star.

893. The most famous comet is Halley's Comet which appears every 75 years. The last time it appeared was on March 8th, 1986 and it will appear again in 2061.

894. Halley's Comet was identified by geophysicist and meteorologist, Edmond Halley. Although the comet had seen countless times before, it was Halley who realised that the sightings of 1456, 1531, 1607, and 1682 were the same comet. He predicted it would appear again in 1758. Although he didn't live to see it (he died in 1742,) his prediction was correct and the comet was named in his honour.

895. 11 comets have been explored by spacecraft.

896. When you look at a comet through a telescope, it looks like a hairy ball with a

fuzzy tail. Because of this, they were originally called Hairy Stars.

897. The ball of the comet is known as the Coma, which is Latin for "hair."

898. A comet's tail can be over seven million miles long.

899. Despite the staggering size of a comet's tail, they have incredibly low density. Most of the tail only has a few hundred atoms per cubic centimetre. If that sounds too sciency for you to understand, the air you breathe is a quadrillion times denser.

900. Sometimes, a comet's tail can split into two.

901. On November 12th 2014, the Rosetta space probe was launched towards the awfully named Churyumov Gerasimenko comet. Rosetta sent in a lander module called Philae. Philae is the first lander to successfully touch the surface of a comet.

Rosetta continued to orbit around Churyumov Gerasimenko for 18 months, photographing and measuring the comet. It even managed to find where Philae ended up.

When Rosetta ran out of fuel for manoeuvring, mission controllers decided to gently crash land the probe onto the comet.

902. Churyumov Gerasimenko looks like it is made from two smaller comets stuck together. Some astronomers say the comet resembles a duck.

903. When Halley's Comet was about to pass Earth in 1910, some people were so scared it would destroy the world, that they bought Anti-Comet pills. Also, Anti-Comet pills were a thing back then.

Black Holes

904. When a star is over 2.8 times the mass of the Sun, it will turn into a Black Hole when it reaches to the end of its life cycle.

905. A Black Hole absorbs everything. I don't mean rocks and stuff. It also devours time so it becomes indistinguishable from regular time. If a person fell into a Black Hole, that person would perceive it normally but to an observer, the fall would take an eternity.

906. Contrary to what diagrams show, Black Holes are not funnel shaped, nor are they flat. Black Holes are spherical.

907. Black Holes are not black; they're invisible. The "black hole" appearance is from the light that the Black Hole is devouring.

908. When a Black Hole devours matter, the matter elongates before being torn apart on an atomic level. If you were sucked into a Black Hole, you would become thinner than a hair but would stretch out so you were about a mile long. This phenomenon is called spaghettification.

909. Nobody truly knows what happens to the matter that a Black Hole absorbs.

910. Some scientists believe there might be an anti-Black Hole (known as a White Hole) that spews out all the matter that a Black Hole has sucked in. However, most astronomers believe the existence of White Holes is highly unlikely.

911. There are approximately 100 million Black Holes in our galaxy.

912. Stephen Hawking developed a theory in university that stated that the universe was created from a Black Hole. Although this isn't true, his teachers found the concept fascinating.

913. Many people believe that once an object is trapped in a Black Hole's gravity field, there's nothing it can do to escape it. Even Stephen Hawking believed this. But it's not that simple.

There is an area around a Black Hole called the event horizon. This is the "no turning back" region. Once you are close enough to a Black Hole, you're going into it and there's nothing you can do to stop it. Stephen Hawking realized that Black Holes can emit particles if they appear in pairs right on the event horizon – one of the pair falls in and the other escapes.

914. Some theorists believe that Black Holes are a type of star that astronomers don't fully understand called a Magnetic Eternally Collapsing Object (MECO.)

915. A normal-sized Black Hole has ten times the mass of the Sun.

916. A normal-sized Black Hole is called a Stellarmass.

917. The most massive Black Hole ever discovered has 17 billion times more mass than the Sun.

918. Some Black Holes are not stationary and travel throughout the universe sucking in all types of matter. These are called Wandering Black Holes and can travel over three million miles per hour. No one knows what causes these types of Black Holes to travel. You know what's scarier? There's one in our galaxy.

919. Most people assume Black Holes work like cosmic vacuum cleaners but this is an oversimplification. Hypothetically, if the Sun turned into a Black Hole, the Earth wouldn't get sucked in. In fact, our planet would continue to orbit it exactly the same way as it orbits the Sun. The Earth doesn't orbit the Sun because it's a star; it orbits it because it has a lot of gravity.

Black Holes also have a lot of gravity. So the Earth would revolve around this Black Hole without devastating consequences for the planet. It only becomes a problem when an object gets too close to the Black Hole. That's when the Black Hole makes you alllllllll kinds of dead.

920. Researchers from the Astronomic Observatory of the Universitat de Valencia have recently witnessed lightning storms near Black Holes.

921. A Supermassive Black Hole (SMBH) is a Black Hole that is dramatically larger than a regular Black Hole. A SMBH can be billions of times heavier than an average-sized star.

922. Scientists witnessed a SMBH devour a star for the first time in June 2018. This Black Hole, which resides in the Arp 299 galaxy, was seen "belching" the remains of the star at a quarter the speed of light.

923. A SMBH exists in the centre of nearly every galaxy in the universe. It is believed that they are vital in the formation of galaxies.

924. The first SMBH was found in our own galaxy on February 15^{th,} 1974 by Bruce Balick and Robert Brown.

925. The SMBH in our galaxy is about 4.3 million times the mass of the Sun.

926. Wi-Fi was invented to detect Black Holes.

The Solar System

927. The Solar System is 4.568 billion years old.

928. The Solar System diagram that you have seen since childhood is not even remotely accurate. That diagram was created for convenience, not accuracy. Most diagrams will show the planets lined up. This only happens every few millennia.

Also, this diagram shows the planets too close to each other. If the planets were as close to each other as the diagram shows, each planet's moons would collide into one other.

Also, Pluto wouldn't be visible on any diagram of the Solar System. Pluto is so far away, that if you had to scale the Sun so it was the size of a beach ball, Pluto would be 1.5 miles away.

929. Because Pluto is no longer classified as a planet, you would assume that there are eight planets in our Solar System. However, it's easier to say that there are eight "official" planets in our System. Scientists keep finding new planets but they can't keep

updating the Solar System diagram as it would become too complex.

Scientists have found planets beyond Pluto like Sedna, Xena, and many others.

930. Beyond Neptune is the Kuiper Belt. This region contains trillions of asteroids and comets. It is similar to the Asteroid Belt except its 20 times wider and 200 times bigger.

931. The IAU (International Astronomical Union) is responsible for naming new moons. Before the IAU was formed, moons were named after the astronomers who discovered them.

The Milky Way

932. Our galaxy, The Milky Way, is about 100,000 light years across.

933. There are between 200-400 billion stars in our galaxy.

934. Our neighbouring galaxy is Andromeda. It has over a trillion stars and is twice the size of the Milky Way.

935. Our galaxy will merge with the Andromeda galaxy in about four billion years.

936. On a clear night, you can see stars up to 19 quadrillion miles away. That's 19,000,000,000,000,000 miles!

937. Astronomers recognise 88 constellations in the sky.

938. The Milky Way moves at 1.6 million miles per hour.

939. If the Milky Way was the size of the United States, the Solar System would be the size of a quarter.

940. Any photograph of the "Milky Way" galaxy is actually of a different galaxy called Messier 74. It's impossible to take a picture of the Milky Way since it's 100,000 light years across... and we are inside of it.

941. There is a force about 200 million light years away that is pulling the Milky Way and other galaxies towards it. It is unknown what this force is but it is nicknamed The Great Attractor.

942. There is a part of the universe called the Zone of Avoidance (ZOA.) The ZOA is the part of the universe we can't see because the Milky Way is in the way. The ZOA obscures 20% of the night sky.

The Big Bang

943. 13.8 billion years ago, a very, very tiny particle started to expand at near the speed of light. Within the first split-second of this expansion, this particle went from being far smaller than an atom to the size of a grapefruit. That doesn't sound that big but that's because it's almost impossible to comprehend how small atoms are. To quote John Green, (who created the YouTube channels, Big History and CrashCourse,) "In much less than the blink of an eye, if it were originally the size of a tennis ball, it would have inflated to over 90 billion light years across."

This sudden inflation is commonly known as The Big Bang.

944. According to scientists, space-time was created at the Big Bang. This means that nothing (including time) existed before the Big Bang. So the Big Bang was literally the first thing to ever happen in existence.

945. When the Big Bang occurred, the universe was 100,000,000,000,000,000,000,000,000,000,000 degrees Celsius. That's 100 nonillion

degrees.

946. Although many people use the Big Bang Theory as an argument to prove God doesn't exist, the theory was concocted by a priest and physicist called Georges Lemaitre. He devised the theory to show that science and religion could co-exist and help understand one another. Sadly, his peers dismissed his theories. Even Einstein thought his theory was baseless.

However, in 2013, the Big Bang was proven to be a fact, not a theory.

947. Lemaitre's theory was called The Primeval Atom. Scientists called it The Big Bang to mock him.

948. Before the Big Bang became a popular concept, scientists believed that the universe was in a never-changing state. They believed it couldn't grow or shrink; only maintain its size.

They also believed that the universe had always existed and never "started."

949. About ten seconds after the universe was created, the laws of the universe like

gravity and electromagnetism were set in stone.

950. When the universe was created, matter and antimatter were formed. If matter and antimatter collide, they annihilate each other from existence. When the Big Bang occurred, 99.9999999% of all matter and antimatter was destroyed. This means that the 0.0000001% of matter that survived the Big Bang makes up the universe of today.

951. Within three minutes of the Big Bang, nuclei and atoms were created.

952. If you turn an old television to a channel without a signal, you should hear static. That sound is cosmic radiation from the Big Bang. Now that television and radios have made the transition to digital, it's getting harder to hear this incredible (but annoying) sound.

953. Robert Wilson and Arno Penzias won a Nobel Prize in Physics for their research that accidentally validated the Big Bang theory. They were working on a totally different project to detect very faint radio reflections, and were trying to get rid of some interference in their equipment, even going

to the extent of removing some roosting pigeons. Eventually, they realised the noise was extragalactic. At the same time, another research group predicted that the Big Bang would have left a background of microwave noise, which was exactly what Wilson and Penzias had detected.

Nebulae

954. Nebula is Latin for "cloud."

955. A nebula is a cloud of soot made up of helium, hydrogen, and other gases. They often form beautiful and colourful shapes.

956. Most astronomical terms are very boring but nebulae have wonderful names such as the Pillars of Earth, the Cat's Eye Nebula, the Horsehead Nebula, and my personal favourite; the God's Eye. (It's the cover of this book.) The God's Eye is so beautiful that it often appears lists called Top 10 Pictures That Look Photoshopped But They Aren't. It is usually ranked at #1.

957. The first nebula was discovered by Charles Messier in 1764.

958. The Andromeda Galaxy was originally thought to be a nebula.

959. Most nebulae are hundreds of light years in diameter. However, 99.9999999999999% of a nebula is completely empty.

960. The Orion Nebula is the only nebula that can be seen with the naked eye. It is below Orion's Belt and is often mistaken for a star.

961. By observing nebulae, astronomers can watch new Solar Systems forming.

962. The Boomerang Nebulae is the coldest natural place in the universe measuring -272.15 degrees Celsius. That's less than 1 Kelvin (less than one degree from absolute zero.)

The Universe

963. The universe is beige. That's not a joke. After astronomers studied 200,000 galaxies, they condensed and compressed the colour of every galaxy and the colour always came out looking beige. Since beige isn't a particularly cool colour, cosmologists tried to change the name of the colour to "cosmic latte." It didn't catch on.

964. Galaxies formed 850 million years after the Big Bang.

965. A gas cloud in the Virgo constellation called RXJ1347 measures 300 million degrees, which is the hottest temperature in the known universe. That's 20 times hotter than the Sun's core.

966. The most common unit of measurement in space is a light year. One light year equals the distance it would take light to travel in one year. Since light is the fastest thing in the universe, (186,000 miles per second,) this is the most logical way to measure vast distances when trying to comprehend the size of the universe.

967. One light year is 5.9 trillion miles. That's 5,900,000,000,000 miles.

968. In 1922, the furthest humanity could see with telescopes was 100,000 light years away.

In 2012, the observable universe was 93 billion light years. That's 548,700,000,000,000,000,000,000,000 miles across. That means that the universe was calculated to be 548.7 septillion miles wide in 2012. However, it would be bigger now because it grows every second.

969. The distances between a planet and its Sun is usually measured in Astronomical Units (AU.) One AU is the average distance between the Earth and the Sun (93 million miles.) So Earth is 1AU from the Sun. Mercury is 0.39AU from the Sun. Neptune is 30AU from the Sun.

970. Although light years are used for measuring celestial bodies, it is not the largest unit of length. A parsec is the longest unit that we use. One parsec is 3.26 light-years. That's 19 trillion miles.

971. In 2011, astronomers found water...in space... just floating around. The water was in a gigantic vapour cloud near a Black Hole. This cloud (known as The Reservoir) had 140 trillion times more water than all the oceans of Earth.

972. It would take 225 million years to walk a light year.

973. A supercluster is a neighbourhood full of hundreds of galaxies.

974. Our galaxy exists with many other galaxies inside the Laniakea supercluster (although many sources inaccurately call it the Virgo supercluster.)

975. The Laniakea supercluster is 520 million light years across.

976. PGC 1000714 is considered to be the oddest galaxy. Although 0.1% of galaxies have rings, this galaxy is the only one ever discovered to have two rings made of countless stars.

977. In Star Trek, Spock lives on the planet, Vulcan. In the Star Trek universe, Vulcan

resides in the orbit of the star, 40 Eridani A. So far, astronomers have yet to find a planet orbiting that star. But we can still dream.

978. There could be as many as 500 billion galaxies in the universe.

979. Each galaxy in the universe could contain as many as a billion stars.

980. Johannes Kepler is one of the most significant people in astronomy. He is considered a founder of modern astronomy and formed the three laws of planetary motion –

i) A planet orbits the Sun in an ellipse with the Sun at one focus

ii) A ray directed from the Sun to a planet sweeps out equal areas in equal times.

iii) The square of the period of a planet's orbit is proportional to the cube of that planet's semimajor axis; the constant of proportionality is the same for all planets.

981. An apogee is when an object that orbits the Earth is at its furthest point. An apoapsis

is when an object is at its furthest from a celestial body apart from the Sun or Earth.

982. A void is a region of space that has absolutely no matter. It is also known as a vacuum. The largest void in the universe is 1.8 billion light years across.

983. To understand ANYTHING about space, astronomers had to understand light. The simplest way to describe light is this – light is an energy wave made from electric and magnetic fields.

984. The largest structure in the known universe is the Hercules-Corona Borealis Wall. It is ten billion light years across. It is made of a vast number of superclusters of stars.

985. Explosions are possible in space but they are not as visually impressive as movies show. Explosions need oxygen. Space doesn't have any. Explosions make a deafening sound. Sound can't carry in space.

986. The universe reacts as if there is a form of matter in it that we cannot see or detect. Although scientists can see this mysterious

matter changing the gravity of galaxies they cannot detect this matter directly. This matter is known as dark matter.

987. Only 4% of the universe is visible. The rest seems to be composed of the enigmatic dark matter.

988. A galaxy called DF2 was discovered in 2018. It is the only galaxy ever observed that does not contain dark matter.

989. Dark energy is negative gravity that plays a role in the acceleration in the expansion of the universe. To date, nobody truly understands how it works.

990. The universe is not circular or spherical as many imagine. It is hyperboloid shaped. In laymen's terms, the universe is shaped like a horse saddle.

991. If you were exposed to space, it would be incredibly painful and freezing but your head or eyes wouldn't explode as many movies suggest nor would you die instantly. Instead, you would die from a lack of oxygen within four minutes.

992. We can never see how big the universe is because we can only observe where light has reached us.

993. 5% of the universe is made of atoms.

994. Isaac Newton used classical physics to understand the universe.

Albert Einstein used relativistic physics to calculate how the universe worked.

Werner Heisenberg used quantum physics to try and understand what knits reality together.

These sets of physics largely ignore each other and even contradict each other. Many scientists wish there was some theory that could connect these three concepts together. This is called the Grand Unification Theory. It may not exist but it's a nice idea.

995. The universe has been expanding since the dawn of time. Scientists assumed that the expansion will decrease over time and eventually stop.

However, studies show that its expansion is accelerating. There is no proof to show that this expansion will be slowing down anytime soon.

996. There are several theories about how the universe might end. One of them is known as The Big Crumble. This theory suggests that the more the universe expands, the more it will stretch itself too far until it cracks open.

997. The Big Crunch is a theory that the universe will eventually stop expanding and instead, start to shrink. According to this theory, the universe would end with everything collapsing into each other. It has been backed up by Einstein's Theory of Relativity.

998. The Oscillating Universe Theory is the idea that if the universe were to collapse into itself as the Big Crunch Theory suggests, the force would be so powerful, that it would create a new universe. Some have suggested that this is how our universe got started. For all we know, our universe could be the millionth time the universe was created. But it still doesn't explain how the very first universe was formed.

999. In 100 trillion years, all the hydrogen of the universe will be exhausted, ending the life of every star.

1000. All matter in the universe will become liquefied in 100 vigintillion years (100,000,000,000,000,000,000,000,000,000,000,000,000,000,000,000,000,000,000,000,000,000,000,000 years.)

Printed in Great Britain
by Amazon